The Unobtrusive Researcher

A guide to methods

Allan Kellehear

Allen & Unwin

For Jan

First published in 1993

Allen & Unwin Pty Ltd
9 Atchison Street, St Leonards, NSW 2065 Australia

National Library of Australia Cataloguing-in-Publication entry:

Kellehear, Allan.
The unobtrusive researcher.

Bibliography.
Includes index.
ISBN 1 86373 513 5.

1. Social sciences—Research. I. Title.

300.72

Set in 10/11.5 pt Garamond by DOCUPRO, Sydney
Printed by SRM Production Services Sdn Bhd, Malaysia

10 9 8 7 6 5 4 3 2 1

Contents

Preface

This book has been written for the growing number of people who believe that there is, or that there should be, more to social research than either surveys or in-depth interviews. Much valuable insight can be gained about ourselves and the lives that we lead by simply listening and watching both systematically and with care. Furthermore, a significant amount of this work can be conducted without either engaging with or disturbing the activity of other people. In other words, much of this research is unobtrusive.

For the unobtrusive researcher, then, the serious study of the 'life and times' of humanity can be found in the thousands of examples and products of our activities: behaviour, dress, organisational spaces, household objects, photographs, books, 'official' statistics, personal diaries, film, music, food and countless other things. By using our ears and eyes, or extending these with cameras, computers and our imagination, the unobtrusive researcher is able to observe some very telling features of our social world—so many of which are regularly overlooked in the common rush to conduct interviews.

The study of archives, physical objects or facial expressions have each become the specialised interest of separate and diverse academic disciplines. And, although archaeologists have a great eye for the study of material culture and historians have a good many archival skills to teach us, few swap notes with one another. This is because few practitioners attempt to familiarise people outside their discipline with their methods and sources. This state of academic affairs is our collective loss and, in the pursuit of a broader vision of social research, a loss no longer worth tolerating. In a world that is becoming increasingly more complex and ambiguous, contemporary social research efforts seem overwhelmed by the task which confronts them. The reliance on single methods, however useful they may have been in the past, is no longer adequate for the task ahead.

In that context, I hope that this book will serve a number of purposes. First, it may introduce readers to a range of sources and methods of social research that people might use to complement their usual research skills and techniques. For example, researchers whose main response to social questions is to reach for a questionnaire might also be persuaded to see the merits of undertaking an analysis of existing sources or to conduct a period of observation.

Second, the book is also designed to revise the reader's view of social research. 'Real' researchers are not simply or solely interviewers and pollsters. Valuable, surprising, provocative and critical research work can also involve working alone in libraries or at computer terminals or sitting in front of a television set. Culture is all around us. Taking time to study some of this material can offer rewarding social insights, contribute to policy formulation or social criticism and generate fertile theories or hypotheses for further testing or debate.

Third, for those who are broadly familiar with the variety of unobtrusive methods of social research but are unsure of the practical steps involved, this book provides introductory suggestions for action and further reading. Rather than simply reviewing studies of material culture or audio-visual records, each chapter offers some practical guidance on how to begin. The suggested reading is designed to take the reader further into both the theory and practice of these methods.

Finally, this book is also designed to be read by people without a 'research methods' background. People who come from a variety of education and training backgrounds but who have not made a formal study of 'doing research' will be introduced to basic research jargon, principles of research design and data analysis. These parts of the book will act as baseline preliminaries to the wider discussion about unobtrusive methods.

The book is organised around eight central chapters. The first defines and discusses unobtrusive methods. Early literature devoted to this style of research is examined with some critical commentary which places this book in that context. This chapter also introduces some basic research terms which are fundamental to any discussion of methods and their efficacy. Chapters two and three review the main principles of research design and analysis. Quantitative and qualitative approaches to research are discussed and their various analytical strategies reviewed. Content, thematic and semiotic analysis is discussed vis-a-vis positivist, ethnographic and post-structuralist research designs.

Chapter four introduces library and archival work. Chapter five extends this interest into the analysis of photographs, film and music. Chapter six discusses research into modern material culture—physical objects, settings and traces. Chapter seven outlines the technical issues involved in simple non-participant observation as a method. The final chapter introduces readers to the world of cameras (still and video) and the adventures of computer modelling. After an initial review of illustrative studies, each chapter discusses the advantages and disadvantages of the source or methods of unobtrusive research. Each chapter also provides practical suggestions for beginning work and further reading and also identifies possible ethical problems with the various methods.

There are not many books similar to this one. Part of the reason for the dearth of guides for the unobtrusive researcher is that anyone attempting this task is confronted with a daunting interdisciplinary project. One must venture away from the safety and surety of one's own discipline, in my case sociology, and attempt to understand and communicate, at a basic level, a diversity of methodological approaches. These approaches are drawn from the research efforts of disciplines which ordinarily have little connection with one another: musicology and archaeology; ethology and cinema studies; history and computer technology; photography and sociology; anthropology and semiotics.

A task such as this could not be achieved without the kind patience, assistance and co-operation of many people from those disciplines. Despite the reservations that some have had about the wisdom of attempting so much in so simple a way, I have been encouraged by the unanimous belief of those colleagues that the attempt itself is nevertheless worthwhile and useful. Not everyone will think that my efforts have been successful but the following people have made major contributions to whatever value lies in these pages.

A number of people provided me with information, discussion opportunities or reading lists in their expert areas. John Goldlust

(Sociology, La Trobe) provided these opportunities for my foray into music; Joan Beaumont (History, Deakin) discussed her experiences with archival work; Derek Colquhoun (Education, Deakin) provided information on his work in secondary analysis; Evan Willis (Sociology, La Trobe) acted as a kind of unpaid research assistant, ferreting out items of interest in newspapers or journals which caught his eye; Shaw Tan (Education, La Trobe) spent much time with me discussing the technical side of photography; Mike McCarthy (Administration, La Trobe) provided me with technical information about copyright; Peter Beilharz (Sociology, La Trobe) was his usual patient self in our several discussions about post-structuralism; Ray Duplain (History, Deakin) discussed registeries and archive work with me and helped with access to some of the photos in this book; Charles Watson (Computer Science and Engineering, La Trobe) thought long and hard in our discussions about the ethics of computer modelling. In the absence of literature in this area the section in this book is a result of his thoughtful suggestions; Brian Taylor and Kathy More (Preston General Cemetery) provided my methods class and me with engaging and often surprising information about cemeteries in Australia. I should also like to thank my 'MDS' methods class of 1992 for its constructive reactions to some of the material in these pages.

Other people kindly read individual chapters and provided helpful suggestions and criticism. Fiona Mackie (Sociology, La Trobe) and John Wiltshire (English, La Trobe) read the chapters on design and pattern recognition. I particularly thank them for their feedback on my introductory explanations of post-structuralism and semiotics. Dennis Warren (Library, Monash), Gwen Scott (Library, Deakin), and Eva Fisch (Library, La Trobe) provided many useful suggestions for the chapter on the written record. Rob Walker (Education, Deakin), Ina Bertrand (Media Studies, La Trobe) and Jeff Pressing (Music, La Trobe) provided critical and useful feedback on the audio-visual chapter. Chris Gosden (Archaeology, La Trobe) guided my reading and writing of the chapter on material culture. His interest and support for a sociologist 'digging around' archaeological terrain is most appreciated. Albert Gomes (Anthropology, La Trobe) provided similar encouragement and work for the observation chapter and Bob Powell (Computing Services, La Trobe) provided invaluable feedback on the hardware/software chapter.

Finally, four friends read the whole manuscript, a true act of friendship at a time when everyone in academia is flat to the boards with work demands. Jan Fook (Social Work, La Trobe) and Jeanne

Daly, Evan Willis and David de Vaus (Sociology, La Trobe) gave their time generously to do me this service. I would also like to acknowledge the careful and patient work of Therese Lennox, Beth Robertson and Glenis Massey as well as Elaine Young (Sociology, La Trobe) who typed the manuscript.

I also acknowledge the following people and agencies for permission to reproduce copyright material: Routledge publishers to reproduce from Ken Plummer (1983) *Documents of Life*, George Allen & Unwin; E.F. Thompkins to reproduce his letter to the Higher Education Supplement in *The Australian*; Professor David Myers (Managing Director of the Thailand–Australia Foundation Ltd of the University of Central Queensland) to reproduce his letter to *The Australian*; Kaz Cooke from *The Age* to reproduce 'Nun's Story' and reply; the *Geelong Advertiser* to reproduce the photos in this book; Michael Leunig from *The Age* to reproduce his poem 'The Wall'; The BBC Enterprises Limited to reproduce the extract from *The Complete Yes Minister* edited by Jonathan Lynn and Antony Jay; Christopher Murphy to reproduce an extract from his regular newspaper column in the *Sun-Herald*. I am also grateful to Echo Fields, and to Anne-Marie Schwirtlich from the Australian Society of Archivists, for permission to adapt their tables of information into the simpler ones that appear in this book. While I have taken every care to contact and acknowledge copyright, both the publisher and I apologise for any infringement which may have inadvertently occurred.

1

Unobtrusive methods: an introduction

There is today, in social science circles, a simple and persistent belief that knowledge about people is available simply by asking. We ask people about themselves, and they tell us. Either we ask them a series of questions in a survey or we have a discussion with them in a structured, semi-structured or 'unstructured' interview. In any case, the assumption is that important 'truths' about people are best gained through talk—a sometimes direct, sometimes subtle, interrogation of experience, attitude or belief. Because the least engaging talk is often seen as superficial it is also seen as the least valid (although the most reliable). Questionnaire surveys, then, are often viewed as the most blunt instrument to record the complexity of human drama. Although often using fewer people, in-depth interviews and the detailed analysis of their 'texts' are seen to penetrate more deeply and sensitively into the subtle world of social and personal meaning. Debate circles, and then becomes tautology. How important is personal meaning if this is not broadly representative? How important can community representativeness be if the responses themselves are hopelessly unrepresentative of the participating individuals? All of this debate comes under the guise of the strengths and limitations of the qualitative versus the quantitative approaches to methods of social research. Yet, as I write this I am increasingly aware that, by and large, we are really speaking about one broadly similar approach to social research. We are writing and debating issues about the information we gain from others when we talk or write to them.

About the only thing that is certain about this method is that it is popular. So, in sociology at least, surveys and interviews tend to be the favoured style of empirical investigation. But this was not always so for sociology and, furthermore, it is not the case for the other social sciences investigating the nature of human culture.

Emile Durkheim, Max Weber and Karl Marx, just to name three of the most well-known figures of sociology's early beginnings, spent most of their time in libraries. Durkheim's archival work with suicide statistics is still debated and celebrated even today, despite its several methodological flaws (Douglas, 1967). History has a long tradition of this type of archival work. Anthropology and education have long favoured the combination of interviewing and extensive periods of observation and informant networking. Both geography and archaeology have interrogated physical objects as evidence for their understanding of culture. In communication and media studies, students of culture have examined art, cinema and music, an extension of the general humanities interest in creative literature. And, although most methodologists in these social sciences in general, and in sociology in particular, will advocate the combined use of these methods, actual teaching practice reveals otherwise. Most disciplines train their own with their tried, true and favourite few methods. It is therefore unusual to find an historian with a questionnaire, just as it raises eyebrows to see a sociologist measuring the height of tombstones in a local cemetery. And so with these circumstances and habits as background, this book is designed to explore other methods of studying human culture which, when *combined* with other methods, may increase our confidence about what we think we are learning about human beings.

In sociology, one commonly finds a token lecture, or book chapter in a methods text, devoted to methods other than surveys or perhaps observations. This chapter or single lecture is usually entitled 'unobtrusive measures' or 'methods'. Sometimes 'non-reactive measures' is the favoured title. These titles refer to social science methods which do not disturb the social environment. They are methods which do not involve talking with people. They are a collection of all the other ways one may learn about human beings and their social world without interrupting them to ask questions. In this chapter I will provide a brief outline of what is meant by the phrase 'unobtrusive methods'. I will then rehearse the advantages and disadvantages of these methods in general. Before proceeding to the following chapters on design and analysis I will spend the final part of this chapter introducing some basic but important concepts in the field of research methods.

Unobtrusive methods

In 1966, Eugene Webb, Donald Campbell, Richard Schwartz and Lee Sechrest published a book entitled *Unobtrusive Measures: Non-Reactive Research in the Social Sciences.* A witty and clever book, it introduced readers to the study of physical traces, archival work, simple observation and the use of hardware. 'Chapter nine' contained only Cardinal Newman's epitaph 'From symbols and shadows to the truth', while the previous 'chapter' contained a single paragraph quote from a statistician imploring us to use 'all available weapons of attack'. The chapter on physical traces gave an overview of the many ways that physical objects can suggest forms of human activity: the wear of floor tiles around museum exhibits as indicators of popular exhibits; the setting of car radio dials as indicators of favourite stations; the wear on library books and rub and fold marks in their pages; content analysis of household or institutional garbage; content analysis of toilet graffiti; gender differences in the shape or height or inscription on tombstones; and so on.

The chapters on archival work begin with the example of a tombstone inscription—an example, I think, more befitting the category of physical traces then 'archives'. However, the review describes the analysis of actuarial (births, deaths and marriage) records; political and judicial records including content analysis of political speeches; government records (power failures, parking meter collections, weather records); media, such as newspapers, and also hospital medical records. Beyond these official sources Webb and his colleagues also discuss the private record: diaries; sales records; advertising; letters; junk mail; suicide notes; and so on.

The chapters on observation outlined studies of hair length, clothes, shoe styles, jewellery, houses, body movement expressions, superstitious behaviour (of baseball players), personal space and public location of furniture, car behaviour, conversation. There is discussion of participant and non-participant observation and also 'contrived' observation, meaning the use of hidden hardware such as cameras, tape recorders, one-way mirrors, 'electric eye' and pressure-measuring devices. In this section, experimental manipulation is also viewed as non-reactive as long as the manipulation is not seen by the subject. The Australian social psychologist Stephen Bochner (1979) also views the manipulative experiment as a non-obtrusive measure. Bochner has made 'wrong number' phone calls to assess 'helpfulness'; has staged fake collapses on trains to assess the same; staged shoplifting to assess shopper reporting behaviour and so on.

However, Webb *et al.* and Bochner emphasise that these types of experiments can be classified as 'non-reactive' or 'unobtrusive' because the people in them are unaware that they are part of an experiment. This 'ignorance' means that their reactions will be natural and therefore not threaten the validity of the findings. This is a very dated view of social research, one which assumes no ethical dilemma in involving people in an experiment without their permission. It also assumes that, providing subjects remain unaware, people going about their ordinary business can be freely side-lined into an experiment that only the experimenter thinks is important. This is an ethically questionable set of attitudes and practices. If people are to be manipulated in some way, permission should be sought because the results may be adverse or embarrassing. For example, a fake collapse may induce a real one in someone shocked by the fake incident. Testing the reporting behaviour of shoppers to a shoplifting event may create quite a deal of emotional turmoil and anxiety both in those who report and in those who choose not to report. Second, lack of awareness in subjects may indeed be unobtrusive for the purposes of validity but it is nevertheless intrusive for those people because their activities have been disturbed by the researcher's activities. The method is therefore *socially* as well as *ethically* intrusive. The greater concern for the effect of reactivity on the research findings and the lower importance attached to social intrusiveness led Webb and his colleagues to drop 'unobtrusive measures' from the title in the second edition of their book. 'Non-reactive' was included in the new title and it assumed greater emphasis.

The happy disregard for the feelings and time of 'subjects' in manipulative experiments conducted in real life settings is a reflection of the minimal level of interest in ethics in recent social science methods texts. Ironically, in the same volume of collected essays as Bochner's description of manipulative experiments (Sechrest, 1979), J.W. Berry in another chapter declined to include these as unobtrusive measures.

Although *Unobtrusive Measures* was a successful book (it had sold 125 000 copies up to 1979) (Sechrest, 1979), the use of these methods has remained modest. The original work was then re-issued in a more recent edition (Webb *et al.*, 1981), but the major problem with both this and the earlier work was the out-dated techniques. Like the earlier discussion concerning ethics and the manipulative

Unobtrusive methods *include*:

- written and audio-visual records
- material culture (physical objects, settings and traces)
- simple observations
- hardware techniques, for example camera, videos etc.

Unobtrusive methods *do not include*:

- interviews, questionnaires
- manipulative experiments
- tests, for example, psychometric tests

experiment, *Unobtrusive Measures* did not include recent developments in computer technology, grounded theory and post-structuralist modifications to content analysis, nor the recent interest and revival of the camera and video. The renewal of interest in cultural studies and its semiotic methods for analysing music, art and film were also either not considered or were considered too briefly by Webb and his colleagues, even into the early 1980s.

For newcomers to these methods, the other problem is the lack of practical guidance in the design of research in general and the execution of unobtrusive measurement in particular. The relationship between the technique and how this might complement lines of thinking about a subject or how this might test or generate theory was largely assumed. Unobtrusive methods were discussed separately from the context of the research enterprise as an investigation which occurs partly in the head, partly in the library and partly in the field. Only the field was given prominence and, especially for undergraduates, this was a disembodied presentation of the research experience. Apart from these problems of history and audience, unobtrusive measurement, like any other set of methods, has its strengths and weaknesses.

General advantages and disadvantages

The advantages of unobtrusive research have been well summarised by Rathje (1979) and Babbie (1989). First, unobtrusive measures tend to assess *actual* behaviour as opposed to *self-reported* behaviour. One of the major sources of error in questionnaires and interviews is believing what respondents say they do or do not believe. The use of unobtrusive methods enables researchers to literally see for themselves. Second, unobtrusive measures are usually safe, both for researchers and other people. Observations, if made discreetly, are harmless and non-disturbing to others; archival and physical trace examination does not even involve physical contact or proximity to other people. For researchers, door-to-door interviewers do not have to deal with territorial dogs or irate respondents, if the information they are seeking can be collected in unobtrusive ways.

Third, unobtrusive methods, because they do not disrupt others, are easily repeatable. This enables re-checking of findings and allows questions of reliability and validity to be re-examined by others. Because of the non-disruptive nature of these methods, the fourth advantage is that people do not react to the researcher. Observations can be discreet and non-involving. As Berry (1979) points out, although some observers may be obtrusive, this observer effect erodes after time.

Fifth, access is not usually a problem. Because researchers rarely need the co-operation of others, research access is much easier. In interviews, each potential respondent must be given an explanation of the research, its possible benefits and harm, and permission must then be sought. For archival work, however complicated the procedure, this may only need to be undertaken once. For simple observations, this procedure may be dispensed with altogether. The study of books or tombstones similarly may require only courtesy permission sought only once.

The sixth advantage is that unobtrusive research is usually inexpensive. Observations, archives and physical trace work are simple, and the major expense is time and a record book, compared with typing, printing and copying expenses for surveys or statistical data analysis, and time and skill in processing survey results.

Finally, because unobtrusive methods are so non-disruptive, inexpensive, accessible and safe, they are ideal for longitudinal study designs—those that follow activities over a period of time. This means, for example, that political or demographic records can be examined over years or decades or that household garbage can be sampled over several weeks instead of just the once, as in cross-sectional studies.

The disadvantages of unobtrusive measures are that the original record, especially for archival sources, may itself be distorted either to hide information or to create a different impression to an outside reader. Not everyone reveals absolutely *all* in a personal diary (in

The advantages of unobtrusive methods are:

- the study of *actual* rather than *reported* behaviour
- safety
- repeatability
- non-disruptive, non-reactive
- easy accessibility
- inexpensive
- good source of longitudinal data

case siblings or parents read it, or journalists steal it); not all basic statistics are 'basic', for example since suicide was regarded as a serious sin by Catholics, many causes of death give another 'cause' on death certificates.

Second, unobtrusive methods are dogged by the usual emic/etic problems. This means that interpretation of physical traces or observations may be from the point of view of the stranger, or outsider (etic) and therefore may fail to grasp important in-group meanings (emic).

Third, intervening variables may also distort data. Rathje (1979) gives the example of garbage analysis which does not take into account recycling practices or hobbyists who sort their garbage for glass, bottles, pop sticks, milk cartons and paper. Cross-sectional work, whether these be trace analyses or interviews, all suffer from similar sources of error. Cross-sectional studies are one-off 'slices of life' and are therefore prone to oversimplification. Sechrest and Phillips (1979) document other intervening variables—the graffiti artist who records a non-existent romance or the alcoholic who deposits his bottles in his/her neighbour's rubbish bin.

Other disadvantages relate to common sources of error such as selective recording of observational data. Certain objects and relations may more likely be recorded by observers with different interests, biases and backgrounds. A male observer of women may notice different features of interaction, mannerisms and dress than a female observer of women, and vice versa. Other sources of error may come from over-reliance on single methods. The research methodological wisdom for surveys and ethnographies is also true for unobtrusive methods—multiple methods are best. Finally, unobtrusive methods (unlike surveys) are limited in the potential areas open to interrogation. Verbal methods, such as interviews, have 'an ability to reach into all content areas' (Webb *et al.*, 1966, p. 181), but this broadness is often difficult to match with unobtrusive method, however imaginative and creative one might be.

The disadvantages of unobtrusive methods are:

- distortion of original record
- decontextualising (emic/etic)
- intervening variables
- selective recording
- single method over-reliance
- limited application range

The above list of advantages and disadvantages in the use of unobtrusive measures highlights or underscores a deeper concern of all researchers. This is the problem of reliability, validity and ethics. As a preliminary and as background to the following chapters I will summarise some of these issues as they reflect generally on social science methods.

An introduction to some basic research terms

In an ideal and artificial way, we can speak about two types of social research, one described as '*theoretical*' and the other type often called '*empirical*. Theoretical research is often library-based research which primarily uses books and articles. The aim of the research is to discuss a theoretical idea, an explanation, through the use of logical reasoning. In this way, certain explanations are modified, challenged or extended by providing a new or different perspective. However, there is no attempt to test this analysis or argument against first-hand collected evidence.

Empirical research is meant to extend beyond books and articles alone so that information (sometimes referred to as 'data') is gathered from other sources—usually people. Anthropologists conduct empirical research by entering villages or cities and watching and talking to people. Sociologists may also engage in this kind of empirical research or they may simply distribute a survey questionnaire. Social psychologists also use surveys or, alternatively, they may conduct social experiments either in laboratories or '*in situ*' ones as Bochner (1979) has done with shoppers or train commuters.

Empirical work does not confine itself to simply interacting with people in one form or another and recording their responses. Empirical work broadly means to test or derive an explanation from active enquiry. This enquiry is commonly through surveys of one sort or another but may also include most of the unobtrusive methods we have mentioned. So research which takes as its data source observations, physical traces and artifactual material such as film, art or music can all be called empirical work. Archival work is sometimes viewed as empirical or historical/theoretical. Because archival work often relies on second-hand information rather than on data gathered directly, this type of work is often viewed as simply historical (Woodland, 1979, p. 66).

All social research involves *methods*, that is, ways of proceeding and collecting information. This ranges from simple procedures such as borrowing and reading books to more complicated empirical procedures such as participant observation. Each method requires that one or a number of *techniques* be mastered in order to apply the method competently. Techniques are the detailed ways, or steps, involved in applying the method. So, for example, if interviewing is the chosen method, the application of this method will require certain technical skills. These techniques range from arranging the order of questions, gaining respondent trust and rapport, ensuring that questions are not leading or loaded in meaning to the technical problem of coding and statistical analysis of responses. An unobtrusive example of the method–technique idea may be the simple observation method. For this method, certain techniques such as coding and recording observations, and deciding on sampling frames are part of the successful execution of the method.

The concern with successful and appropriate methods for the research task has to do with every researcher's deeper concern for the issues of reliability and validity. *Reliability*, or a reliable method, refers to a method which, if used by others in similar conditions to the original research, will actually turn up the *same* or highly similar results. There is little point in skipping through a cemetery and, after a quick browse through the epitaphs, concluding that the graves of men enjoy longer epitaphs than women if the next person who skips through arrives at the opposite conclusion. Research must be systematic, organised and disciplined so that those who re-test or re-examine your source of data will arrive at similar findings. This is one important source of confidence for you and others that your method is a good, reliable one.

The second major source of confidence concerns the validity of the method. *Validity*, or the issue of how valid a method is, refers to how well a method measures what it claims to measure. Consider the example of social class. There are many possible methods of measuring social class. Some may argue that asking people to tell you their occupation is a reliable and valid measure of social class. Individuals are usually quite happy to disclose their occupation, and occupational prestige is a valid measure of social class (Daniel, 1983). This is because occupations are tied closely to other indicators of class such as income and education. On the other hand, some people might regard simply observing the car that people usually drive as a valid measure of social class. However, many working-class people drive expensive cars and many middle-class people drive inexpensive cars. Quite a few business people rent cars that they could not afford to buy while more than a few wealthy people drive inexpensive and old vehicles. Many people do not own cars and some people drive other people's cars, particularly young people driving their parents' cars. So observation of people in their cars is a much less valid measure of social class than asking people their occupations. However, *combining* the observations with interviews may actually increase the validity of the findings because one method may turn up findings which can be explored by the other.

The issue of validity is not intrinsic to a method because this validity can only be assessed in relation to the problem to be researched. Observations of cars are probably less valid sources of data than interviews with their drivers when it comes to assessing social class.

Sampling is another issue in social research. If you wish to understand the daily life of Swiss migrants who own coffee shops in Melbourne, it would be desirable to study them all, if their population is less than fifty. However, if you wish to study the daily life of Melbourne's university students you will be obliged to study a sample because the population in which you are interested numbers thousands. How do you know if your sample truly represents the rest of the population of university students? This is a question commonly at the centre of sampling concerns. The ideal is to procure a sample that has all the main features of the population and does not over- or under-represent one or more of these features. There must be so many men and so many women; there should be so many school leavers and so many mature-aged students included; there will need to be a certain percentage of the sample who are from professional, science- and arts-based courses; so many from part-time and full-time enrolments; so many born in

Australia and others who were born overseas; so many undergraduates and so many postgraduates; and so on.

Sampling does not only apply to people. It also applies to observations, when and whom one observes. Sampling also applies to physical traces from graffiti to tombstones. In archival work, a sampling frame or system may be useful in examining the contents of diaries and letters or government records, particularly if the volume of these is enormous.

The main research terms are:

- theoretical—research based on other research, reason and second-hand information.
- empirical—research based on first-hand experience and designed to test or generate theory.
- methods—ways of proceeding and collecting information.
- techniques—steps involved in the execution of the method.
- reliability—the dependability or confidence one has that if a method is used by others in similar circumstances that they will arrive at similar findings.
- validity—the precision or confidence one has that the method measures what it claims to measure.
- population—the total number of people, physical traces, pages, words etc. (called 'units') that you may wish to know about or research.
- sample—a systematically collected part or specimen of the population which may give you some informative ideas about the population as a whole.

Finally, all good research is concerned, not simply with issues of how confident we are that our findings will be believed (theory vs. empirical work; how sound or appropriate the methods; validity vs. reliability; how careful the sampling) but also with *ethics*. Ethics should be understood to be a normal part of any concern with method. Because methods are about ways of proceeding, ethics concerns itself with the most socially responsible way of doing this. Research ethics refers to the responsibility that researchers have towards each other, the people who are being researched, and the wider society which supports that research.

Researchers are expected to consider the *safety* and welfare of those people who participate in the research. The community expects that no harm should come to people who are the subject of research. Furthermore, researchers should protect the confidences and identity of those they research, they should guard the *privacy* of those they study. In this respect, peoples' lives should not be unduly intruded upon. The community also rightly expects that any research which involves certain groups or that may have findings which publicly reflect on them in some way, seek the subject's *permission* or *consent.*

Cheating is also a practice which is not welcome in social research. One should not deceive the people who are the subject of the study; one should not falsify or misrepresent the findings of research; one should not use the theoretical or empirical work of others without acknowledgment; and, finally, one should not claim, through authorship, research to which one did not actively contribute.

Another ethical issue concerns *the use* to which research may be put. Some research is acquired through significant expense of time and convenience of respondents or subjects and then the findings simply rest on library shelves. Much thesis work for higher degrees suffers this fate, a practice which is arguably irresponsible. Research findings should be disseminated in professional and/or community circles if at all possible. If this is not practicable then out of courtesy those who participated in the research might be informed of the findings for their effort.

Other research findings may be put to negative use by governments or by other sectional community interests. Groups involved in advertising may use findings to exploit the vulnerability of some groups, while other agencies may use social research findings to persecute dissident or marginal groups. All research has social implications. Splitting the atom or documenting the plight of the rural poor are findings useful to someone. The ethical concerns are: who are these people and to what use will they put these findings and, if I anticipate possible harm, what steps should I take to protect my respondents?

Webb and his colleagues (1966) in their original work devoted two out of 225 pages to a discussion of ethics. Their conclusion was that so few people could agree on a comprehensive criteria for ethical research that a separate work would be needed. And with that assertion the four authors dust off their responsibilities to discuss research and ethics as integrated concerns and activities. In the second edition (Webb *et al.*, 1981), this oversight is addressed with a specific chapter but a certain type of hedging around the issues continues. The authors, now numbering five, rehearse the pros and cons of privacy, consent, confidentiality and protection of subjects vis-a-vis covert observations, archives and other non-reactive methods. Issues of cheating and the social and political uses of findings are ignored perhaps because the authors feel that these matters have more to do with self and/or peer regulation. Consent, privacy, confidentiality and protection directly concern respondents or subjects and it is this 'direct relationship', focusing on 'the other', which drives their interpretation of research ethics. This means, of course, that only one side of the moral equation in research is being explored.

Furthermore, Webb and his colleagues maintain that ethical criteria for research are difficult to arrive at and suggest only that researchers keep a watchful eye on state, national and institutional developments in the area of research ethics policy. No practical suggestions for researchers, especially for novices, are made.

Just as methods must be chosen with a research problem in mind, ethics should also be discussed in this context. I have said elsewhere (Kellehear, 1989) that ethics cannot simply imply the use of broad checklists. Research ethics often arise from, and are suggested by, the research endeavour itself. In this context, ethical dilemmas and issues should be discussed with co-researchers at the time of research planning as part of the design. Ethical problems may be identified at this stage and ways to overcome them planned. As the research progresses from design to field, further problems may present themselves that were not anticipated. As with similar methodological problems, these should be shared and discussed with others. If one is a lone researcher, one should still always discuss potential ethical issues with other researchers. Ethical concerns are social. Ethical guidelines are developed because we desire to avoid harming others. This social concern is about accountability and it therefore reaches its optimal consideration only if discussed with others.

When the findings have been analysed and are part of a final report, thesis or article, dissemination and implications of dissemination are important topics of ethical discussion. I am not referring here to submissions which may be taken to institutional committees—sometimes referred to as 'ethics committees'. Where these exist, researchers may or even must avail themselves of their scrutiny and advice. And even though I think that the proliferation of such formal devices has its merits, the disadvantage is that this may abrogate the researcher from the responsibility of seeing ethics as part of the *ongoing process* of research. Too often ethics committees monitor only the first stage of research, the proposal. Too often the major ethical dilemmas arise in later parts of the research, the field and the report. Consequently, all researchers should monitor and discuss these issues among themselves as ongoing concerns and practical problems.

Ethical concerns refer to:

- privacy of subjects
- consent of subjects
- confidentiality of subjects
- protection from harm of subjects
- cheating—avoiding deception of subjects, colleagues, community
- negative use of research by self and others (including choice of research topic and empowering or disempowering publishing practices for the researched or co-authors)

As ethics committees focus on research subjects (animals or humans), these committees often give the false impression that research ethics apply mainly to that particular interface. That focus ignores other ethics such as cheating with results, copyright infringement, choosing not to publish one's findings etc. These and many other examples are not covered by that interface. The stated priorities of ethics committees can give the false impression that ethics is about 'what we do to others' rather than the wider moral and social responsibilities of simply *being a researcher*. Ethics is always about fair and honest dealing, whether towards an active participant or with colleagues or State agencies, or towards owners of a diary, for example. I will spend a little time at the end of each subsequent chapter outlining some of the areas of possible ethical concern when we cover the various types of unobtrusive research.

And, finally, ethical problems occur over the three phases of the research process and discussion with others in each of these is the best way of self-monitoring.

The three phases of research are:

- research design
- in the field
- write up and public reporting

Recommended reading

Babbie, E. (1969), *The Practice of Social Research*, 5th edn, Wadsworth, Belmont, CA (especially part 5 including appendices A and B)

Bulmer, M. (1984), *Sociological Research Methods*, Macmillan, London (see especially part 3)

Douglas, J.D. (1967), *The Social Meanings of Suicide*, Princeton University Press, Princeton, New Jersey

Kellehear, A. (1989), 'Ethics and social research', in J. Perry (1989), *Doing Fieldwork: Eight Personal Accounts of Social Research*, Deakin University Press, Geelong

Kirk, J. and Miller, M.L. (1986), *Reliability and Validity in Qualitative Research*, Sage, Newbury Park

Sechrest, L. (1979), *Unobtrusive Measures Today*, Jossey-Bass, San Francisco

Singleton, R., Straits, B.C., Straits, M.M. and McAllister, R.J. (1988), *Approaches to Social Research*, Oxford University Press, New York (especially chapter 12)

Webb, E., Campbell, D.T., Schwartz, R.D. and Sechrest, L. (1966), *Unobtrusive Measures: Non-Reactive Research in the Social Sciences*, Rand McNally & Co., Chicago

Webb, E., Campbell, D.T., Schwartz, R.D., Sechrest, L. and Grove, J.G. (1981), *Non-Reactive Measures in the Social Sciences*, 2nd edn, Houghton Mifflin Company, Boston

2

Principles of research design

Most books or articles on research design begin with abstract principles and end in one of two ways: either the advice concludes with an example or, alternatively, with a short discussion of how research reports should be presented. The abstract principles introduce difficult ideas such as 'conceptual definitions', 'problem analysis', 'paradigm selection' and so on. The example of design toward the end of that kind of discussion is designed to shore up all the readers who got lost in the preceding discussion. Advice which takes this approach moves from the abstract to the concrete tasks of design. I think this process of explanation should be reversed. It is more helpful and enabling to move from the more easily understood aspects of research design to the more abstract. So, I will begin with the main ways in which research is presented and then look at the rationale and reasons behind the design.

Every research report has subtle differences in presentation which are due to personal style and the idiosyncratic demands of the data collected. However, the main influence on research reports which determines their presentation is the author's methodological preference. By methodological preference I mean whether researchers start out with a theory which they wish to test (hypothetico–deductive) or whether they conduct research to develop theory (ethnographic–inductive research).

Hypothetico–deductive design

The hypothetico–deductive design is so called because it tests hypotheses (hypothetico) and it does this by testing a general theory on a particular sample of cases. It moves, therefore, from the general idea/theory/hypotheses to the particular (deductive) sample of a study. The first section of this kind of research report is the *literature review*. In the literature review researchers identify and discuss the main literature which is relevant to the research endeavour. If the research is about 'car theft in urban areas' then the literature will contain all or the main studies on car theft in urban areas. The discussion will assess the strengths and weaknesses in former studies and summarise the main findings and explanations (theories). The extensiveness and depth of this literature review will depend on whether the research is for an article, a report, or for a thesis or book. If the literature review is written for a journal article, the review is usually brief. The review will contextualise the proposed or actual research and highlight the main gap to which the current research responds. For a book or a thesis, a literature review may be anything up to 20 000 words of extensive and exhaustive critique and evaluation. This is because authors are attempting to appear to master the whole area of their interest so as to impress colleagues or examiners. In any case, the literature review is always an evaluative task which poses questions and identifies gaps or omissions in the study of a certain area.

The next major section in an hypothetico–deductive report is the 'theoretical framework'. The theoretical framework is the outline of an explanation that you believe will address the gap or omission in the literature. It is an idea or set of ideas which, if supported by some empirical evidence, will allow the research area to proceed further than it has hitherto. In my research with the dying (Kellehear, 1990) my review of the literature revealed two problems with the sociology of death and dying. First, I did not agree that we are a 'death-denying' society as many did and do still believe. Second, the sociology of death and dying had studied death fairly extensively but not the behaviour of the dying. If I could show that a dying role still existed in modern society then that would be strong criticism of the idea of social denial. This is because dying, to be considered a public form of behaviour, would need to be seen as a set of *mutual* expectations and exchanges. How would I test this idea and what might a dying role be? I combed the historical and anthropological literature on dying and I noted that the behaviour of those dying in most times and places had several reoccuring features in common.

Five features of what many have called 'the good death' were:

awareness of dying (from self or others), adjustments to illness and altered social relations, preparation for death, disengaging from work and saying good-bye. The 'good death' therefore, was my 'theoretical framework', my set of ideas which would add to our understanding of dying and help challenge the notion that we are a 'death-denying society'. These five features needed identification and discussion with any supporting literature brought to bear on the ideas themselves. Furthermore, these ideas would serve two further practical functions. First, they would serve as themes or discussion areas in my interviews and, second, they would serve as hypotheses.

As hypotheses, each of these features of the 'good death' had to be specified in more precise terms. Did farewells really take place between dying people and their intimates? For most or for only a few? If most of the dying made farewells then the hypothesis here is considered proven and one feature of the modern 'good death' is established, at least for this study. Theoretical frameworks are *a priori* organising ideas, that is, ideas developed *before* the empirical encounter with the world.

The next section in a hypothetico–deductive report is methods. The methods section identifies the main methods and techniques employed to collect the data for or against the theory. This may involve describing the choices of interviews, observations or archival work. Many studies employ only one method, such as interviews, while others combine several. The choice must be explained and preferably justified in terms of past research and/or current resources. The simple choice of methods is not the only issue discussed. The finer points of research design are also rehearsed and justified here: studies might be cross-sectional, that is a once-only study in one timeframe (for example, a survey of voting attitudes in February 1993); they might also be longitudinal—a series of studies of the same people, concerning the same issues, over a period of time (for example, a voting attitude survey of the same people conducted annually over five years). Some research designs might be experimental, that is testing individual reactions to an object or activity that you, as researcher, are manipulating. In this context, less obvious, but no less important, are design considerations which focus on issues such as social bias in question style, sample selection or choice of methods. This is particularly important if it can be shown that the design might distort or discriminate against one or more social groups. (For an example of this see Eichler's 1988 discussion of gender bias in research design.) In this context, the limitation of the method and design should also be discussed here along with any other ethical considerations.

The section on method is followed by one on results. Traditionally, this section is the one which contains the statistics. I say 'traditionally' because the hypothetico–deductive style has usually been home to the positivist approach to research. The questionnaire–measurement interests of this approach have usually favoured statistical techniques and it is the results section where these are rehearsed and corralled. In a short article the results section can be quite major. Here, hypotheses are supported or discarded, interesting 'levels of significance' identified, and frequencies which allude to other interesting patterns of behaviour or belief are described. For observational studies, a quick overview of one's observations is also contained in this section.

Finally, a discussion section appears. Discussion sections of a report, thesis or article are designed to link back to the literature review and theoretical framework sections of the report. How right were we after all? How wrong were we? Where did we go wrong? How complete is our explanation now that the evidence from this

The hypothetico–deductive research report comprises:

- *literature review*
  - —reviews previous studies and theories, evaluates
  - —identifies gaps and omissions of previous work, locates present study
- *theoretical framework*
  - —outlines a theory or explanation which addresses past omissions
  - —provides detailed theoretical exposition of these ideas
  - —suggests hypotheses
- *methods*
  - —identifies methods appropriate to the theory
  - —justifies and explains the choice of methods
  - —ethical discussion
- *results*
  - —description of statistical analysis, observations or other measurements
- *discussion*
  - —attempts to explain the findings
  - —refers back to literature and theory section
  - —suggestions for future work

particular study is all in? What findings need further explanation? How to account for these findings in the light of the theory? Does the theory need modification and in what ways? How might those modifications themselves seek empirical validation and support in further work?

The above presentation format is also the basic outline of research design. However, research design is not simply about the major steps to research writing. Design is also about what method or methods are best chosen for the research question you have in mind. If your main interest is in generalising from your data, you will need information which is reliable and representative. Often, your sample will be large—counting 200 tombstones or collecting 200 questionnaires. If you are primarily interested in meaning and experience, you might choose to confine yourself to ten long interviews or a lengthy, thematic analysis of personal diaries or journals.

The hypothetico–deductive rationale for research comprises:

- read first (literature review)
- get an idea (theoretical framework)
- go out (methods)
- test it (results)
- see if you were right in the first place (discussion)

However, not everyone thinks that this is the best way to undertake social research. There is a belief that the hypothetico–deductive approach imposes a set of meanings on social phenomena. It also excludes from study ideas which are not part of the original theory. In these senses, quantitative/theory-testing/deductive approaches develop an 'outsider's view' of the social phenomena that are being studied (etic viewpoint).

Those who feel that these criticisms should be incorporated in the research design tend to design their work differently.

The ethnographic–inductive design

The ethnographic–inductive design is so called because it has often been favoured by anthropologists in their fieldwork. Anthropologists develop a picture of society through a multitude or combination of methods. These primarily include observations of one sort or another but also include some interviews, the use of informants, and the study of physical objects, geography and ecosystems. The 'ethnographic method', then, is less a method than an approach to analysing and portraying a social system. The ethnographer attempts to understand the commonsense meanings and experiences of the participants of a social system. This is sometimes referred to as attempting to understand the social system from 'the insider's' point of view (emic viewpoint). From a study of this system of social life, one attempts to develop an explanation about the development, maintenance and salience of certain social processes. In this sense, one moves from the particular case (the study) to the general social theory, an inductive movement of thought which is sometimes referred to as 'grounded theory'.

These studies also begin with a literature review. The aim of the literature review is not simply to gain an impression of theoretical and empirical omissions alone, although this is not unimportant. The task of the literature review is also to develop a 'sense of place'. By this I mean to develop a social picture of the setting and its people. Whether you are interested in the experiences of inmates in a prison or the social life of the Fijian-born Indian, the review of the literature should help convey past academic attempts to tell this story. The literature review then, for inductive research designs, is not only an evaluation of past literature, it is also the background to the culture studied, with imperfections of the literature noted and discussed.

The issue of correspondence is important to both hypothetico–deductive and ethnographic–inductive studies. The idea of correspondence is actually quite simple and can be encapsulated in the following question: how well does my social description correspond with the reality? In ethnographic–inductive designs, researchers often take the view that the theory, the explanations, the connection between action and interpretation, should be suggested by the social system itself. They are interested in immersing themselves in the social system, or the accounts of that system, and developing the theory by observing how patterns of meaning emerge from the social practices and beliefs of those they are studying. These emerging patterns are the social stimuli upon which a theory is constructed.

It follows from these beliefs that the literature review section is not followed by a theoretical framework section. Instead, ethnographic description follows the literature review. In this section the setting, the people and their way of life are described in considerable detail. Since the foundation of any theory will be developed on the basis of this description, that description must be thorough. Like a novelist, observations and interviews must be attentive, receptive, facilitative and guided only in informal and informed ways. The richness of this description forms the basis of this section which may be entitled 'ethnographic description'.

The section which follows the ethnographic description is usually the discussion section. Note that, in addition to the omission of a formal theoretical framework section, no results section exists either. The ethnographic description section is the 'equivalent' of a positivist's results section. The discussion is usually quite lengthy and sometimes may include several subsections. Whatever the particular style, the idea is nevertheless quite similar. The discussion should attempt to construct a social portrayal of the people studied vis-a-vis the past understanding of them as contained in the academic literature.

The abstract social processes, symbols and structures which underlie the ethnography should now be brought to the fore of the study in this section. The connections between the various abstract explanations should be justified by their close links with the narrative account or observations. Sometimes the discussion section is not sharply divided from the ethnography but rather is a self-reflective and critical extension of it. The important point to realise is that the important analytic energies which drive the analysis, and hence the written presentation, are description (ethnography) and explanation (theory).

A section describing methods may occur in two places. Either reflections and descriptions of method may be placed in an appendix or, alternatively, this discussion may occur after the literature review. If the methodological discussion occurs after the literature review it may not do so as a formal section. Sometimes a section entitled something like 'Background to the study' might contain among other topics, discussion of method. In this section one may describe the physical setting, its demographic, geographic and historical background as it is known. Also contained in this section are reflections on the steps taken and difficulties encountered in entering the field situation. Reflections on ethics and a review of the methods employed in data collection may be part of this preliminary exercise in self-reflection.

The above format or presentation is an expression of a research design which seeks to develop explanation from ideas and experi-

The ethnographic–inductive research report comprises:

- *literature review*
  - —reviews previous studies and theories, evaluates
  - —identifies gaps and omissions of previous work, locates
  - —provides historical perspective to understanding
- *background to the study*
  - —describes physical, demographic, geographic and historical setting
  - —describes entry to field
  - —describes methods and ethical dilemmas
- *ethnographic description*
  - —describes social life, sometimes organised around themes or case studies
- *discussion*
  - —develops theories, interpretations for the social symbols and processes described
  - —integrates or takes issue with previous literature in this area

ences suggested by the social system itself rather than simply from the academic's discourse. It does not ignore that discourse but rather uses this as an instrument for orienting the researcher to past ways of understanding. Once oriented, the researcher then attempts to enter the world of the 'other' by actively cultivating an empathy for that world. This empathy together with field notes are ways in which the researcher will later construct a formal theory of social life.

The ethnographic–inductive rationale for research implies:

- reading first (literature review)
- gaining experience, participate, listen, record (ethnographic description)
- describing the theoretical implications of what you saw/heard (discussion)
- explaining where you were and how you went about your task of understanding (background to the study)

The above organisation/designs for writing and research are illustrated in a general perusal of the academic social science literature. From that perusal you will see that psychologists and quantitative oriented sociologists favour the hypothetico–deductive format. Anthropologists and qualitative sociologists tend to favour the ethnographic–inductive presentation. You might examine the following examples for further reflection in this area.

Some examples of the hypothetico–deductive design:

de Vaus, D.A. (1990), *Surveys in Social Research*, Allen & Unwin, London, chapter 18

Goode, E. (1982), 'Multiple drug use among marijuana smokers', in G. Rose, *Deciphering Sociological Research*, Macmillan, London, pp. 192–206

Heaven, P.C.L. (1991), 'Voting intention and the two value model: a further investigation', *Australian Journal of Psychology*, vol. 43 no. 2, pp. 75–7

Some examples of ethnographic–inductive design:

Cleghorn, P.L. (1981), 'The community store', in R.A. Gould and M.D. Schiffer (eds), *Modern Material Culture: The Archaeology of Us*, Academic Press, New York, pp. 197–212

Klofas, J.M. and Cutshall, C.R. (1985), 'The social archaeology of a juvenile facility', *Qualitative Sociology*, vol. 8 no. 4, pp. 368–87

Minichiello, V., Aroni, R., Timewell, E. and Alexander, L. (1990), *In-depth Interviewing: Researching People*, Longman Cheshire, Melbourne, appendix A.

Note carefully that the above examples of design and presentation are most commonly found in academic and professional journals. They are also the most common styles used in theses and formal reports. These are all areas of heavy peer review and regulation and this is the major reason for the high level of conformity and low level of imaginative and interesting presentation. When you begin to look beyond these specific instances and, say, browse academic books published by commercial publishers, things become rather more interesting and presentation of work takes on a diverse and less formalised appearance.

Both theoretical and empirical work of this type tends to shape itself around the main issues, themes or aspects of an argument. If an argument is ordered into five or six stages, or applies itself to three or four examples, then the chapters will follow or reflect that organisation. Sometimes a central argument in a book is not discernible so that perhaps the aim of the author/s is merely to show, or demonstrate, certain ways of seeing or analysing issues. The chapters will order themselves along those issues with preliminary and concluding chapters.

More often than not though, studies (empirical or theoretical) have something to say—they wish to persuade. Each author attempts to work out the best way to do this through the organisation of the book, report or article. The story of the study must unfold in parts so that the reader is kept interested, all the while following the logical development of the author's thought within the work's organisation.

Many books do not have a discussion of literature separate to that of method. Many books omit a methods chapter altogether. A book that relies heavily on archival methods may simply state that and move on to the more important task of developing the insights from the study. Literature reviewed for the purpose of the study or argument might be integrated throughout a book. This is a favoured style of 'post-structuralist' researchers (also known by the equally ambiguous label 'post-modernist').

Post-structuralist writers, partly because they wish to avoid the authorial voice of 'the expert' and partly because they have no use for self-conscious 'scientific' presentations, often write as if they were telling a story. They are 'up front' with their attempts to tell a social story, explain an account of things, in a different way. They wish to reveal, in the re-examination of a piece of social reality, that which might not be obvious to the casual observer. They may use several examples to make their point or they may engage with several central writers to do this. Each presentation will reflect the idiosyncratic style and choices of each writer.

In these above respects, when referring to the organisation of commercially published books or, more specifically, the style of a certain group of qualitative writers (e.g. post-structuralists) no formulae of presentation can be outlined—almost anything goes! Doing justice to the research topic is confronted by the demands of doing justice to the writing as a literary exercise and the readers as potentially diverse audiences. A good research report or thesis does not make the best read, however clear the organisation of that report is meant to be. A good book, however, is more than a matter of clarity. It is also one of style, engaging persuasion and memorable reflections, and there is no ready design for these goals.

The earlier designs and presentations are themselves designed to cater for the needs of *peer accountability*—so that others can check your review, methods, results and thoughtful or thoughtless logic. Good books, designed for wide readership within and beyond one's discipline, must meet broader and less tangible criteria for 'good' reading. For post-structuralist writers, 'research' *is* a 'reading' of the world, and the task is always persuasion rather than proving.

Their model, therefore, has always emulated the literary, creative model rather than the formalistic, scientific–academic model.

Rationale and problems

As I mentioned earlier the main influences behind the hypothetico–deductive design are the quantitative approaches to the social sciences. These approaches have had a number of philosophical influences which were important in their developing and justifying this approach. Quantitative researchers believe that their research designs are scientific. They, more than other social scientists, emphasise theory construction and concern for the issues of validity and reliability. Modelling themselves on their understanding of the physical sciences, they believe that theory building and testing is the way forward in understanding the world, including human behaviour. Measurement is important to this understanding because measurement provides a strong basis for testing old theories or developing new ones. The world 'out there' is a mystery and the task facing social science is to develop theories to explain it.

A narrow or exclusive focus on measuring things scientifically is usually referred to as 'positivism'. The idea that the world can be understood in terms of relations between a thinking and knowing self and an outside, not-self world is part of the view called 'empiricism'. An empirical view is one that says that the world 'out there' is a reality that can be visited and studied objectively. The aim of quantitative oriented scientists is to establish cause and effect and to examine the logic of the social universe. This concern with the logic and order of the world is sometimes referred to as 'rationalism'.

Philosophical influences behind hypothetico–deductive research are:

- empiricism—I go out into the world
- positivism—with my own theory and measure
- rationalism—a set of cause and effect relations

The ethnographic–inductive format, favoured by the qualitative researchers, is influenced by different philosophic traditions. Qualitative social scientists have strong reservations about imposing a pre-structured theory onto the world. Human beings are not physical objects but, rather, conscious, decision-making and often irrational beings. Order is often unstable and changeable. Cause and effect are artificial concepts which oversimplify complex, continuous processes of metamorphosis and ambiguity. Social science should indeed go out into the world but with only a desire to listen and participate. One must bracket one's former understanding about particular social phenomena and attempt to understand these processes from the point of view of the experiencer. This concern with the insider's point of view is often called phenomenology, while a concern for the way in which people interpret and make sense of their world is called symbolic interaction. The world 'out there' is several 'worlds'. There is less concern for the objective and more concern for how people make and understand their world. Worlds are 'live-in' places and the ethnographer must socially or at least psychologically, try to enter that world. The world is not objective but subjective.

The philosophical influences behind ethnographic–inductive research are:

- empiricism (naturalism)—I go out into the world (into 'habitats')
- phenomenology—and describe from the insider's point of view
- symbolic interaction—the way people interpret the world

Within the qualitative group of researchers there is yet further dissent about how to conduct social research. One group of qualitative researchers does not believe that there is an objective world 'out there'. The world 'out there' is, in fact, a purely social construction. People have 'understandings' about the world which they share with others as stories or narratives. However, these narratives are not simply interpretations about the world because the stories themselves actually constitute the world. In other words, the stories are instrumental in socially and physically shaping the world. The design of the world and the interpretation of the world are one. But the determining influence or energy comes from the dominant narratives.

The narratives, however, are not themselves any more objective than the world that they purport to interpret. On the contrary, the stories people hold about their world—about housing, gender, health, religion or architecture—are built upon certain hidden agendas. The main way in which people understand things reveals other issues or symbols which may be denied, hidden, repressed or in any case effectively silenced. This principle, that culture is a network of narratives or stories built on the hidden, comes partly from psychoanalysis but also partly from literary theory. The views from this particular group of qualitative researchers are sometimes referred to as 'post-structuralist' or 'post-modern' or 'social constructionist' (see Lyotard, 1984). The aim of these researchers is to 'rewrite the narrative' by uncovering ('deconstructing') those hidden but powerful elements. The distinction between self and the world and theory and empirical work is spurious, that is, not what it seems. All social research which is critical and penetrating is deconstructive, unpeeling the variety of human meanings unconsciously inherited by people in the course of their life and socialisation.

The philosophical influences behind post-structuralist research are:

- phenomenology—I describe (uncover, 'deconstruct'), beginning from the native's viewpoint
- symbolic interaction—how the dominant way people interpret the world
- psychoanalysis/semiotics, literary theory—is based upon one or *several* hidden conventions/agendas. These 'rules' behind the conventions have the effect of limiting/suppressing other voices, ideas, knowledge or experiences (for example, women, blacks, alternative health ideas, parapsychology etc.)

The differences between the quantitative and the qualitative research approaches, and between those whose qualitative approach is influenced by traditional anthropological or semiotic sources, is important. In the next chapter we will be analysing the main techniques used in interpreting data collected through the use of unobtrusive methods. The philosophic influences introduced here will also be useful in understanding which techniques are favoured by different researchers and the broad reasons behind those preferences. But before we launch into this technical exposition, now is a good time to have a brief word or two about 'research ownership'.

Ethics and research ownership

Many people write alone. They design a study, collect experiences or data, go into a room somewhere and write all about it. Many theses are written this way. But beyond the writing of the solitary worker, many others work collaboratively in matters of design, data collection and analysis, and finally jointly write the report or publications which stem from the earlier work.

If you work alone on a theory and not see anyone, the work thereby produced is yours (assuming no plagiarism). Usually most people do not work that way, help comes with contacts, access, advice, practical support, data analysis or reading report drafts and so on. Such helpers are acknowledged with a thank-you in the acknowledgments and a box of chockies in private. Here, too, are the people who participated in the study, those interviewed, relatives who own those letters and diaries and so on. Depending on how many of them there actually are, you might want to hold off on the chockies but somewhere, in good research, that reciprocity needs to be respected.

Then there are other people who are actually doing the study with you. These are the people who wrote part of the report or collected a large part of your data or helped in important ways in the actual design ideas. These people are usually co-authors of reports or publications which stem from the work. Some people believe that if you 'pay' a research assistant then they can be asked to do almost anything and should be happy with an acknowledgment. I think this exploitative practice is quality nonsense. A research assistant is a helper with transcription, statistical computation, typing and literature location, search and retrieval. As the task of the assistant becomes more complex and involving, the role of that person becomes one of research associate. This is a person who gathers a substantial amount of the data, who writes a major portion of a report or who contributes major ideas which change the design of the research in important ways.

Everyone has slightly different ideas on research assistants and associates and everyone has differing ideas on what 'major contribution' actually means. But if you are part of a group or team, ensure that your know your role and the credit which will be accorded to you for your work. These matters should *never* be left unclear (see Dunkin, 1992). A major role performed in the research should entitle you to co-authorship of some of the publications. That question should be settled clearly at the beginning of any research collaboration. Anyone who tries to convince you otherwise is trying to sell you something. Don't buy it!

I leave this chapter with a complaint sent to Kaz Cooke in her 'Keep Yourself Nice' column in *The Age* (25 January, 1992). It is signed 'Nun's Story' and it is all too often a 'common story'.

> Dear Kaz,
>
> I am a humble scientist, who does believe she is making a difference to patient care. My problem is that while I come up with the ideas, my medico boss insists on standing in the limelight. I know I should be pious and show humility, but inside my head I keep having very un-nice thoughts. How should I repent?
>
> Nun's Story, Brunswick
>
> Sometimes people present ideas as their own on purpose, and sometimes they have mulled it over long enough in their spacious heads, they become convinced it was their brainwave all along. The various approaches you may employ include saying in front of other people, 'I see you've implemented my idea for . . .' (fill in flash of brilliance here). 'I've been thinking more about it, and it might be even better if. . .' (further insight here). Or you could tell him or her to knock it off. Or you could say, at an appropriate juncture, 'That was my idea. Give it back, give me credit, or I'll start calling you "Junior"!'.

Recommended reading

de Vaus, D.A. (1990), *Surveys in Social Research*, Allen & Unwin Hyman, London, chapters 1 to 3

Dunkin, M. (1992), 'Some dynamics of authorship', *The Australian Universities Review*, vol. 35, no. 1, pp. 43–8

Eichler, M. (1988), *Nonsexist Research Methods: A Practical Guide*, Allen & Unwin, Winchester, Massachusetts (see appendix especially)

Game, A. (1991), *Undoing the Social*, Open University Press, London, chapter 1

Glassner, B. and Moreno, J.D. (1989), *The Qualitative—Quantitative Distinction in the Social Sciences*, Kluwer Academic Publishers, Dordrecht (see especially Robert Fellapa's article in this anthology)

Lyotard, J.F. (1984), *The Postmodern Condition: A Report on Knowledge*, Manchester University Press, Manchester (see especially pp. 9–17 and 79–82)

Minichiello, V. *et al.* (1990), *In-depth Interviewing: Research People*, Longman Cheshire, Melbourne, chapters 2 and 3

Murray, K.D.S. (1992), *The Judgment of Paris: Recent French Theory in a Local Context*, Allen & Unwin, Sydney

Najman, J. (1992), 'Comparing alternative methodologies of social research:

an overview', J. Daly, I. McDonald and E. Willis, *Researching Health Care: Designs, Dilemmas & Disciplines,* Routledge, London

Wadsworth, Y. (1984), *Do It Yourself Social Research,* Victoria Council of Social Services, Collingwood

3

Principles of pattern recognition

There are many techniques for interpreting patterns in data. This is true for patterns in written text form (for example, letters, diaries, interview transcripts, statistics, books); in visual form (for example, photos, film, observations of behaviour or physical objects); and in audio form (for example, music, speech, traffic noise, audience noise). One simple way of introducing the principle of pattern recognition is to think about the common task of note-taking when reading books. You may remember in the last chapter that all research designs begin with some kind of preliminary reading or review of the literature. Reading and noting down one's impressions of various books and articles are rather important preliminaries in the research process.

Of course much of what you take down as 'notes' depends very much on *why* you are reading the text and what you wish to *take* from the reading. In other words, the purpose of your task is the issue behind what determines what you see as being important or unimportant, relevant or irrelevant. This applies to the whole issue of pattern recognition and not just note-taking, that is, a reading of written text material. The task of understanding and of theorising, or developing explanations of phenomena, begins with the researcher being able to discern a recurring pattern in the music, the written word, the behaviours, the sounds, the physical objects.

Some researchers come to these sources with pre-structured categories which they have developed and look for these in the data. Others look at what themes or categories are suggested by the data itself. Or, one might attempt to discern patterns behind the

Ways to 'take notes'

1 Simply summarise one's impressions intuitively.
2 Summarise with a pre-structured view (using one of several article analysis forms which are available, for example, MacQueen, 1973, p. xix; Knop, 1969, pp. 67–71; Runcie, 1980, pp. 21–2).
3 Summarise only features which arbitrarily interest you.
4 Summarise using the author's organisation as the main guideline.
5 'Read between the lines', attempting to summarise the author's intent or hidden agenda.
6 Ignore the task of summary, noting only your view/or the author's 'major points'.
7 Note only the major 'themes'.
8 Note only the argument and its evidence.
9 Note only the paper's omissions or your view of its weaknesses, biases etc.
10 Note only its stylistic qualities, literary devices, or language.

obvious patterns which are suggested by the data. Researchers who come to social phenomena with pre-structured categories which they wish to count or observe tend to be quantitative/positivist workers. Their technical approach is often called 'content analysis' (note-taking styles 2, 3, 6). In the note-taking metaphor, one has a list of topics or issues that one is looking for and one simply applies this to all the readings (2). In this sense, you only summarise or note those issues which interest you (3) and (6).

Researchers who wish to be more inductive and to develop categories from the data which might reflect the culture and its people tend to be the qualitative workers. Their technical style may be described as 'thematic analysis' or 'grounded theoretical analysis' (note-taking styles 4, 6, 7, 8, 10). In the note-taking metaphor once again, one carefully examines the author's organisation and takes notes which reflect this examination (4). The summary is guided by the author's main points (6). The themes which appear to be important to the author, especially through his or her argument, are especially noted (7, 8). Finally, any qualities of style, supporting evidence, methodological practices which are part of the author's work are noted (10). Post-structuralist analysts take this grounded theory approach further by searching for omissions and oppositional symbols which may reveal one or several hidden agendas. The idea here is to uncover the 'rules' or 'mentality' behind a certain set of texts, images, object-relations or behaviours. Often, as mentioned earlier, this is tied to some critical theory of culture, politics, self or the body. The technique, which is rarely given much explicit discussion, is often called a 'semiotic analysis' (note-taking styles 1, 5, 6, 7, 9). In the note-taking metaphor, one summarises intuitively, sensing with your experience and interests what is, or what is not, important in the text (1). There is an attempt to read behind the text, to look for meanings beyond the author's obvious intention. One may note the author's background, qualifications and other work to help in this task (5). During summary one notes the author's view of the major points but beside this one also notes, perhaps in the margins, one's own view of these—your own major points about the author's meanings (6). Manifest and less obvious themes are noted (7). Any omissions, gaps and inconsistencies are noted to help you understand issues of bias, manipulation, hidden agendas, unstated interests and so on (9).

I will now turn to a separate and more detailed introduction to each of these. When reading the following discussion remember that the differences between content, thematic and semiotic styles are not as necessarily clear cut and distinct as they are presented below. These styles are simply ways of seeing and analysing patterns in the social world. I have broken them up into clean categories to make their analytical priorities clearer. In reality though, there can be much overlap in these styles.

Content analysis

Content analysis is the act of developing categories prior to searching for them in the data. Frequently, the category is quantified, that is, the number of times it occurs is counted. In observations one may count grimaces, or sitting, or eye contact, touching, dominating or submissive behaviour and so on. In written form, one may count certain words, phrases or ideas as these appear in the text. Wiseman and Aron (1970) cite the example of a search through women's magazines to see what kinds of heroes are presented. One may choose the magazines, the time period (between 1960–70) and the number of issues to be examined. They suggest hypothetically that the categories representing the heroes could be divided into physical, emotional, social and personality characteristics. For example, physical characteristics might be age, gender, weight/shape, hair, eye and skin colour, scars etc.

For content analysis, only the imagination will limit the kinds of things one may search and measure: values, attitudes, words, number of pages, types of images and sounds, inches of type, minutes of television or film time and so on. Singleton *et al.* (1988, pp. 349–50) outline the basic areas which one might quantify: *time–space measures* (newsprint column inches devoted to a certain topic/s in a newspaper, hours of televised violence, for example); *appearances* (How many television advertisements use male voice-overs? Is the central character in American films white, male and between 30 and 45 years of age, physically well built and educated?); *frequency* how many times does a word or an idea appear in a policy document or textbook or speech?; *intensity*—this involves measures of frequency and importance or prominence in the behaviour, text or objects. In general 'the categories are constructed by the researcher' (Scott, 1990, p. 130) and this is a central characteristic of the technique. Scott (1990) discusses three main requirements which should be met by content analysis so that its conclusions are seen as credible (that is, reliable and valid). These are *comprehensiveness*, which means examining all the relevant sources and not just those which support your own theory. The second requirement is that the categories be *specific and clear*. The categories must not be overlapping. A thorough typology of categories which are independent of each other is important to minimise ambiguity and maximise reliability. The third requirement is that categories be subject to *clear definitions*. According to Scott, the clearer the definition (so that even a computer could allocate the data into them) the better and stronger the reliability.

In the past, content analysis has been quite useful in a number of ways. According to Holsti (1969), content analysis has been useful to:

1 Secure political and military intelligence. This application was used for the purposes of analysing wartime propaganda in the 1940s.
2 Analyse the psychological traits of individuals. This has been particularly useful in biographic work in the use of diaries, letters and other personal documents.
3 Analyse aspects of culture and cultural change by examining documents over long periods of time.
4 Provide legal evidence. Holsti offers the example of a govern-

Practical principles in content analysis

- decide on certain categories (from the theory or questions)
- choose the sample
- select the time period for sampling (if relevant)
- decide on the number of events to be observed, issues to be read, shows/films to be seen
- record the observations systematically (tables can be useful here)

Measures in content analysis

- size of time/space allocations
- simple appearance
- frequencies
- intensity

(Singleton *et al.*, 1988)

Categories used in content analysis

- comprehensive (systematic and unbiased)
- specific and clear (not overlapping or ambiguous)
- supported by precise definitions (for ease of data fit)

(Scott, 1990)

ment standards organisation which monitors television content for bias against minority groups.

5 Assess authorship. Occasionally documents or artwork is discovered but no signature or authorship is known. Detailed content analysis which compares the idiosyncrasies of style with content helps to identify authorship.
6 Measure readability. Hence, a certain judgement is made about topic interests and level of literacy difficulty/complexity and this is applied to set literature or cinema for appropriateness of audiences or readership. Cinema censorship is technically of this type.
7 Analyse the flow of information. For example, comparing news bureau outputs (AAP, Reuters etc.) with news stories as they appear in the newspapers.

Holsti also provides the example of experimental work which supplies a certain piece of information to a group and then documents its distortion by later interviewing others about that information.

Clearly, the number and diversity of applications is some testimony to its popular use and the common view of its advantages. As historian Carney (1972) observes 'you can be sure what you have been looking for and where you have been looking for it'. Second, others are easily able to check your data and your method of classification. Third, a wide range of work can be performed with these types of techniques. Fourth, the influences between data and interpretation are close and explicit. Finally, the systematic nature of the approach allows researchers to control their intuitions and tendencies towards distraction and bias.

The disadvantages, however, are just as numerous and are to be watched for closely. Singleton *et al.* (1988) warn that space–time measures are *gross* measures and may be imprecise in smaller or more subtle social issues and areas. This also applies to appearance measures. Frequency measures make two hazardous assumptions: first, that frequency is a valid indicator of its importance and, second, that each individual count is of equal value or importance. For example, articles on page one of a newspaper may be of greater impact or influence than those on page sixteen.

Scott (1990) warns of etic problems. Because content analysis is often deductive (derived from a researcher's theory or hypotheses), the validity may be a problem. Since the categories and therefore the major meanings are imposed, there is little that may be truly applicable from the researched group's point of view. An outsider's interpretation may be a fiction if there is no way to check with the person or culture one is researching. Also, Wiseman and Aron (1970) warn that records tend to survive if these were thought to be of sufficient importance. Letters or works of art or policies thought trivial some time ago but which may tell us important things now may have been destroyed or not recorded. We must also be aware of the statistical axiom that 'correlation is not causation'. Associations or patterns may reflect other patterns, they may not cause them. In any case, it is a rare content analysis which can technically arbitrate on these type of issues.

Finally, the most frequent criticism of content analysis is that the fetish for frequency makes the technique atomistic. This means that it breaks data into small, decontextualised and hence meaningless fragments, and then reassembles them using the researcher's own framework (as an outsider). Grounded theory and/or semiotic analysis are able to overcome these problems by attempting to develop their theories/understandings by inductively attending to the data and its source.

Thematic analysis (also known as grounded theory or narrative analysis)

Thematic analysis derives much of its approach from the sociologists Barney Glaser and Anselm Strauss. They stress the view that validity is tied to how well a researcher's understanding of a culture parallels that culture's view of itself. The central meanings attached to objects or relations should reflect the beliefs that the insiders hold about these. The analysis may go beyond these meanings but if those meanings are the starting point then they had better be valid ones. Validity here begins with the convergence of researcher and the subject's ideas about the subject's view of the world.

In the case of interviewing, of course, one can check back with the respondents about the themes which emerged from the interview. The researcher can then revise the 'thematic analysis' in the light of the subject's comments and additional input. The inductive approach in thematic analysis means that rarely is a 'theoretical framework' or set of hypotheses used. Instead, the researcher is interested in a topic or set of issues and then approaches an interview or document with these issues in mind. Themes are then sought after as these emerge from the narrative of the interview or written word or behaviours. Let me quote from two workers who are reflecting on this process of identifying themes in their field notes.

Mori Insinger (1991) is a sociologist who interviewed eleven people who had close brushes with death. He was interested in conducting a thematic analysis of the effect of this event on their social networks. He describes below how he undertook this task:

> I analysed the data by first transcribing all interviews. I then individually and thoroughly examined the transcriptions, and coded each part of the interview theme by theme. I then broke down general themes into extensive sub-themes, in an attempt to capture the many aspects of family interaction that may be correlated with the after effects of the near-death experience. Finally I disassembled the interviews, and reorganised them by thematic classification for presentation. (Insinger, 1991, p. 145)

The finer task of discovering or discerning themes and giving those themes names (coding) is given more detailed discussion by Patton (1980) and Viney and Bousefield (1991). I quote some of Patton's (1980, p. 299) practice below.

> I begin by reading through all my field notes or interviews and making comments in the margins or even attaching pieces of paper with staples or paper clips that contain my notions about what I can do with different parts of the data. This is the beginning of organising the data into topics and files. Coming up with topics is like constructing an index for a book or labels for a file system; look at what is there and give it a name, a label. The copy on which these topics and labels are written becomes the indexed copy of the field notes. . .'

When Patton writes about 'topics' he is really talking about the same thing as Insinger who talks about themes.

In one's field notes lay tracts of description of observations; or this may be the text of a diary; or perhaps a series of tombstone inscriptions; or perhaps the words of a series of popular songs. In any case, the task is to break the whole into parts, sections which have 'smaller bits of meaning' in themselves. This process is also known as 'unitising'. While content analysis may refine this down to words, thematic analysis usually looks for ideas in the narrative or the text being examined. Overall, however, it must be remembered that the grounded theory/thematic analysis takes its categories from the data or field notes themselves. Furthermore, the analysis is pushed along or suggested by the priorities and values in these notes. Counting, although of some use in thematic analysis, does not occupy the central place that it does in traditional content analysis. Moreover, *a priori* categories which come from theory tend not to be the main approach. The strong reliability which comes from well-defined coding categories gives any such application a strong impression of validity (Scott, 1990, p. 155). Thematic analysis, on the other hand, is more subjective and interpretive. Thematic analysis which takes the data itself as the orienting stimulus for analysis attempts to overcome etic (outsider's) problems of interpretation by staying close to the emic (insider's) view of the world. Also, thematic analysis does overcome the problem of viewing all events or items as having equal value or importance. Similarly, it also does not accept that frequency is, in itself, a valid or reliable indicator of importance.

Subjective as it is, thematic analysis is more demanding on the personal resources and intellectual art and craft of the individual researcher. What happens if you read the text or describe the observations but can think of nothing to say? What if no theme emerges from the data? A quick perusal of the personal stories recounted in the field work literature (Bell and Encel, 1978; Perry, 1989) reveals that this is not an uncommon problem, particularly for neophyte researchers.

With this brain-block in mind, Miles and Huberman (1984, pp. 215–49) have a series of suggestions to help one discern themes or 'generate meanings' from the data. They make twelve suggestions which I have summarised below into eight points.

The Miles and Huberman approach is not the only way of

> Practical principals in thematic analysis
>
> 1 Count—look for repetition, recurring events/experiences/topics.
> 2 Note themes, patterns—look for underlying similarities between experiences.
> 3 Make metaphors, analogies or symbols for what is happening.
> 4 Check to see if single variables/events/experiences are really several.
> 5 Connect particular events to general ones.
> 6 Note differences and similarities.
> 7 Note triggering, connecting or mediating variables.
> 8 Note if patterns in the data resemble theories/concepts.
>
> (Miles and Huberman, 1984, pp. 215–30)

generating meanings, of course. Kern (1970, pp. 553–61) suggests a less inductive, even more structured approach to data which is drawn from Karl Mannheim's (1936) work.

The Kern/Mannheim approach has a lot more in common with the post-structuralist/ semiotic approach, something I will say more about below.

Many of the above tasks (Miles and Huberman and Kern/Mannheim) may be performed manually by sorting and scribbling on your field notes. At other times, computer programs can be used to perform these tasks. ETHNOGRAPH and an Australian program called NUDIST (Nonnumerical Unstructured Data Indexing, Searching and Theorising) are able to format text, allow segments of text to be identified by a code and retrieved (Richards and Richards, 1991). Such programs can deal with documents, field notes and even photo-images. NUDIST and ETHNOGRAPH can also provide technical

Kern's (1970) suggestions for thematic analysis

1. Select a period or problem to be investigated.
2. Identify a leading thought or idea *and* its opposite.
3. Trace all ideas to one or another category and develop a 'left-overs' category.
4. Analyse all data to see how well or not they fit the leading idea category and its opposite.
5. Attempt to discern the thought styles of the group or classes which are associated with the manifest idea by going beyond the spirit of the leading idea.
6. Begin process again with the left-over categories.

(Summarising Mannheim [1936])

support for problem solving and testing with a variety of in-built programs and retrieval techniques.

For now, let's return to the suggestions by Miles and Huberman. These suggestions indicate that looking for themes in data, with minimal preconceived categories, is a creative, imaginative and time-consuming task. It actually reminds me of my high school days of analysing poetry. One works through the poem almost intuitively, looking as you go, for patterns of style and content which will yield a richness of meaning within the linguistic devices, the music and the metaphors and symbols of the unfolding verse. One reads, and then rereads the text. An idea occurs and you reread with this idea in mind. Sometimes the idea (theme) 'makes sense' and allows you to understand the poem at another level or to connect with other words or symbols. At other times, the idea does not, you feel, gain support from other parts of the poem and you reject the idea and look again for another. The poem or text is combed repeatedly until you arrive at an interpretation that you feel able to defend or support with references to that text, its organisation and symbolic structure. For support in your beliefs about your interpretation you may read other poems or texts from the same author. This 'art' of interpreting literary texts has been given the name of 'hermeneutics' by many people in the humanities. Hermeneutics is simply the art and study of interpretation. It may be of interest to know that this study began in earnest during the Reformation when controversy existed over the 'proper' interpretation of the Bible (Connolly and Keutner, 1988).

That concern over whether a text had a one 'true' interpretation or whether a text had as many interpretations as one might have readers has continued in the humanities until this day. Some believe that texts do have at least one 'decidable' (i.e. true) interpretation because of the author's intention, but others believe that all interpretations vary within the preconceived ideas of the reader's time and culture (Connolly and Keutner, 1988, p. 26). This debate about the status of interpretation has irked those in the more positivist–quantitative traditions. These people feel that this ambiguity undermines the issue of reliability and validity in measurement. Those in hermeneutics who believe that interpretations can be rigorous and subject to evidence and reason disagree. Those in the hermeneutic traditions who believe that interpretations and textual readings are the same things do not care. They argue that issues of reliability and validity are of little concern when the central issues concern the nature of meaning, experience and power. The authority of concepts such as reliability and validity is spurious, that is, not what it seems. Scientific reliability says more about the culture of professional conformity than it does about the pursuit of understanding. And validity in science, as any reading of the history of science will show, is a changeable concept subject to shifts in the culture of the scientific community itself. Those sentiments reflect the attitude of post-structuralists and their techniques for analysing texts. The techniques of pattern recognition used by the post-structuralists are often called semiotic analysis.

Semiotic analysis

In a way, just as thematic analysis is a modified outgrowth of more traditional content analysis so too then, semiotic analysis can be seen as a yet more modified thematic analysis—a more radical version, even.

Semiotics' first home was the humanities, particularly literature and language (linguistic) studies. Similar to hermeneutics, its main concern was the question of how to interpret meaning, particularly in language. The simple view of language is that words refer to objects in the world. One can understand a text by simply adding up the collection of all that those words refer to. But certain scholars, Saussure (1966) and Derrida (1981), argued that texts could only be understood in terms of their internal component parts. Themes and messages in any written form can be understood in terms of their reference to other themes and messages in the text. So far this is similar to a thematic or hermeneutic analysis. However, semiotic analysis often goes a step further. The themes *omitted* or repressed and/or overlayed by other themes are also part of the analysis.

Furthermore, in an effort to go beyond the confines of the text itself, one can call in to service certain historical concepts such as power, class or patriarchy. These ideas draw attention to the social and political context of the text. When these theoretical traditions form part of the interpretation this is often referred to as a *materialist semiotics* (Game, 1991). Post-structuralist researchers transfer this set of techniques, this style of analysis, to social settings, behaviours and images but also to scholarly debate in art, the sciences and the humanities (discourse analysis).

Semiotics today therefore refers not simply to the uncovering of rules, relations and mentality behind written texts but also to social life and all its symbols in general. The influence of psychoanalysis and feminism frequently guides semiotic work to look at the role of the unconscious in the shaping of culture and its many social and aesthetic expressions (and exclusions). In this tradition, it is the hidden, the distorted and the repressed messages and relations which become as important as the obvious and explicit ones. Indeed, the uncovering of this layer of meaning alters our understanding of the other taken-for-granted meanings of life. As Leunig (1992) quipped in an edition of *The Sunday Age*:

> The great dividing wall came down
> But it hasn't gone away;
> It lies flat on the surface now
> To keep what's underneath at bay;
> And what's below our surface
> Has become the 'other side'
> More dreadful than the communists
> Across the old divide.
> Yes the wall's been repositioned
> And its thick and strong and flat;
> To keep us from the awful things
> Swept underneath the mat.

One important difference between a semiotic analysis and a thematic analysis is that semiotic work is not necessarily inductive. Deeley (1990, p. 12) regards semiotic analysis as less a method than simply a 'point of view', technically similar to behaviourism or positivism. He argues that a semiotic analysis really involves following one realisation about hidden meanings and codes to its fullest unfolded implications. As Manning (1987, p. 31) remarks 'Interpretation is fundamental', but commonly this interpretation is led by some model or theoretical framework. This might either be very general (for example, psychoanalysis, Marxist–feminism) or it may be very specific (for example, writers such as Foucault, Lemert, Derrida, Irigaray, Benjamin).

Similar to content analysis, semiotic analysis brings to its observations a set of assumptions and theories; intellectual baggage, if you like, which is unpacked around and by the data collected. Beyond simply bringing the idea of the hidden and repressed to texts and 'deconstructing' (uncovering), post-structuralists are suspicious of the ideas of scientific truth, of social appearance, of the idea of progress and linearity in history and of the singular authoritative voice in professional writing. All these ideas scream out to a post-structuralist that something, indeed many things, are being hidden from view. The preference of scientific, artistic or popular thinking to embrace the idea of unified, harmonious and continuous wholes, whether this be the very idea of an Australian history or the idea of a thing called a discipline of psychology or 'the individual versus society' is rejected. These are the type of structures which post-structuralists wish to forget and go beyond. All reflections reveal. However, reflections which embrace the idea that social objects, relations and experiences have single meanings oversimplify the situation. The essential idea behind semiotic analysis then, is to uncover, to 'deconstruct' or decode. This 'uncovering' process helps reveal the operation of the dominant and powerful codes and practices of the day.

Mounin (1985, pp. 103–4) provides a simple example. Advertising imagery follows certain rules of construction. These might be informed by psychological, aesthetic, graphic and, no doubt, other influences and theories. But, ordinarily, the viewer of these advertisements does not decode these theories, assumptions and strategies, strategies which are supposedly attracting his attention and guiding his or her thoughts and emotions about a particular message or product. Indeed, if the hidden code were analysed one could argue that the intended message would be substantially altered or even rendered impotent. An important requirement of successful advertising is clear reception of a message with minimal or no awareness of the technicalities of the manipulation. If everyone analysed every advertisement, advertising would lose considerable impact and advertisers would lose considerable money. The idea is *not* to think about the hidden but rather to attend to the simple and obvious appearance before you—the ad as is.

Another rather simple example can be drawn from the television series *Yes Minister*. In one scene, a cabinet minister is discussing the problem of 'getting anything done' in office. He is discussing this with a colleague who is presently in opposition. That minister revealed that public servants (such as the Minister's department head, Sir Humphrey) had a technique for stalling ministers from doing anything. However, as seen on separate occasions, that technique was not obvious, indeed, even had a semblance of reasonableness about it. I quote from the actual conversation (Lynn and Jay, 1984, p. 93).

> According to Tom, it's in five stages. I made a note during our conversation, for future reference.
>
> *Stage one*: Humphrey will say that the administration is in its early months and there's an awful lot of other things to get on with. (Tom clearly know his stuff. That is just what Humphrey said to me the day before yesterday.)
>
> *Stage two*: If I persist past stage one, he'll say that he quite appreciates the intention, something clearly ought to be done—but is this the right way to achieve it?
>
> *Stage three*: If I'm still undeterred he will shift his ground from how I do it to *when* I do it, i.e. 'Minister, this is not the time, for all sorts of reasons'.
>
> *Stage four*: Lots of ministers settle for stage three according to Tom. But if not, he will then say that the policy has run into difficulties—technical, political and/or legal. (Legal difficulties are best because they can be made totally incomprehensible and can go on for ever.)
>
> *Stage five*: Finally, because the first four stages have taken up to three years, the last stage is to say that 'we're getting rather near to the run-up to the next general election—so we can't be sure of getting the policy through'. The stages can be made to last three years because at each Sir Humphrey will do absolutely nothing until the Minister chases him. And he assumes, rightly, that the Minister has too much else to do'
>
> (Extract reproduced from *The Complete Yes Minister* edited by Jonathan Lynn and Anthony Jay with the permission of BBC Enterprises Limited.)

This 'deconstruction' of a simple set of relations assumes that the conversations and relations concerning this issue *are not what they seem*. Indeed, as Game (1991, p. 128) reminds us about bosses and secretaries in general, master–slave relations are often not what they seem.

Several other ideas follow from this first idea and support it. The common unifying and taken-for-granted binary assumptions must be overturned to understand beyond superficial appearances. Assumptions about the powerful and powerless, controlled and controllers, harmony and conflict set our cognitive co-ordinates so that we 'see' in conformist and superficial ways. Alan Roughley (1991, p. 22) argues that these binary meanings (he mentions others such as male/female, good/bad, light/dark) are understood preferentially. One term is historically preferred and 'allowed to control it'.

If these preferences are put aside, one soon discovers that fixed meanings give way to meanings that are processual and fluid. Ideas and relations become changeable, associated as they often are with struggle, conflict, uncertainty and ambiguity. How then does one 'put aside' these binary arrangements? How do the post-structuralists see beyond the flashing lights of our binary prejudices?

Roughley (1991) suggests that deconstructing begins with 'marking' the way that any text is structured by pairs of binary meanings. After this, attempt to rearrange your understanding of this pairing. Attempt to swap over your opposites or look for codes which may reveal a more potent or influential set of relations which hide behind the original binary pairing. Who has the power in this scene? He has. Really? Let's assume she has, or assume she has some of the real power some of the time. Now look for it. Who is the 'hero' in this film? That person is. But what if the 'villain' is the 'real' hero? Let's take a closer look at that shall we? Is that really a love song? Might it be something else? What else could it be? For whom are the meanings obvious? Who might they serve? For whom are the other meanings which are so difficult to discern? Why are they so difficult to discern? What does this say about all of us?

Roy Turner (1989) attempts to demystify the process of deconstructing text. The 'text' in this case is his field work among former mental patients. His analysis is to scrutinise certain binary notions such as patients (versus 'the well'), former mental patient (versus never a former mental patient), the Enlightenment notion of social problems (versus normality, wellness, harmonious relations). By overturning these ideas, that is, by subjecting each of these to critical comparison with each other, Turner concludes that former mental patients are just like everyone else. They turn a social designation into an identity and they have problems resuming relationships. Furthermore, the label of 'former mental patient' is neither distinctive nor definitive, unless written up as a research report in precisely that way.

Evan Willis (1986) has attempted to 'deconstruct' RSI (Repetitive Strain Injury) as a social process. I talked to him about his method. He sought and assembled a wide range of documents from library, media and government sources. He was interested in two levels of meaning, the level of obvious appearances and then the underlying level of meaning. In this case, he examined the media literature, the scientific literature and the popular literature to ascertain what was being said and what was missing or de-emphasised. To work out what was missing or being de-emphasised, Evan created a list of all the issues, topics and themes emergent in all the literature. Obviously, however, not all the literature addressed all the issues or themes. Why not? Why did the medicos emphasise different things to the trade unions? Was there another agenda apart from the 'straightforward' one of 'understanding and solving the problem of RSI'.

To help tease out this agenda Evan used work by Karl Figlio (1982) which examined the social and political meaning of illness. He also looked at 'labour-process theory'. This is a tradition of literature which examines the social changes in the workplace that are associated with the introduction of new technologies. Work processes create new technologies and also technologies change workplaces and practices. The issue of control and surveillance of the workplace is important here. How does the debate about the 'reality' or 'myth' of RSI fit into this interesting scheme? Suddenly, the issue of the reality of RSI is bypassed, overturned, given a new alternative set of meanings. Now the RSI debate is about control, surveillance, dehumanisation of the workplace, injuries as mediators of industrial relations.

Game (1991) works in a similar fashion. As mentioned earlier, she was interested in boss (he)–secretary (she) relations. Analysing the transcripts of her interviews she challenged the notion that power, control and reciprocity is all what it seems; '. . .she does not simply take up his positioning of her. She performs her task to keep the job, but not as he imagines her doing so' (p. 128). The relationship is not a closed system but rather locked into broader gender relations. Reciprocity is not reciprocity. This is because the central meanings about each other are not shared. She is not defined by his gaze, attitudes or values and he is not understood by her in his own terms. Conflict, subversiveness and disruption are not events but active, if not socially apparent, elements of the relationship. Similar to Evan Willis, Ann Game sensitises herself to these shifts of meaning in her data by drawing on Irigaray, Lacan, Hegel, Freud and other theorists.

Now, I do not make the claim that all post-structuralists perform

Practical principles for semiotic analysis

1. *Assemble* all available material together, supplying as much information about their cultural and physical context as possible (Manning, 1987).
2. *Identify* the oppositional elements (for example, boss/secretary, scientific opinion/lay opinion, active/passive words).
3. *Compare* the different accounts/elements. As Game and Willis have done, read the stories, accounts, versions 'across and against each other'—a practice sometimes called 'intertextuality' (Game, 1991, p. 48).
4. *Trace* their source beyond their setting (for example, how many 'voices'/ideas in this letter? Whose ideas are they? What vested interest does each party/group have in the successful portrayal/acceptance of this image/'truth'?
5. *Critically apply* theoretical traditions which are sensitive and cognisant of social processes which repress, distort, mystify and oversimplify social relations. The main ones here are Marxist–feminism, psychoanalysis, critical and discourse theory, and post-structuralist writers, for example, Derrida (1981), Foucault (1972), Lyotard (1984), Kristeva (1987).

their semiotic analysis in the above way. My only comment is that if you follow these steps you will produce a modestly robust semiotic analysis which would be quite intellectually respectable. Most semiotic analysis follows these principles in some form or another.

The problems with a semiotic analysis often depend on how complex and elaborate any author takes principles four and five to be. Writers may trace the source and interests behind texts to remote connections or perhaps those connections themselves are poorly made and supported. Many traditional Freudian and Jungian analyses of text suffer from this problem (Singleton *et al.*, 1988, p. 351). The application of post-structural theories may also be uncritical, making the connections drawn open to charges of over-interpretation. As Scott (1990, p. 148) reminds us 'researchers who are studying a text in order to discern its internal meaning are simply one of its many audiences'. Meanings cannot be separated from their audience. I remember watching the television mini-series *The Dismissal*, a series which re-enacted the dismissal of the Whitlam Labor Government by the then Governor-General Sir John Kerr in 1975. My impression was that the series would alienate a lot of Liberal Party followers and supporters since the series was clearly sympathetic to the Labor follower's perspective. I was constantly surprised by meeting so many Liberal supporters who felt exactly as I did but in reverse. They felt the series was quite critical of the Whitlam government and would therefore offend Labor supporters! People come to their readings and viewings with frameworks, often unconscious, which help them select and focus their gaze. Bias is no less a problem for post-structuralists in their pursuit of 'deeper' meanings than those whose search is for the merely manifest.

Other commentators are less restrained in their criticism of the post-structuralist's use of semiotics. Some of this anger has recently spilled over into the Higher Education Supplement of *The Australian* newspaper. Eric Thompkins (1991) from South Australia commented that:

> The meaning i.e. the use of a word in one context, in one language game—one socially accepted, tacitly rule-regulated usage—bears no necessary relation to its meaning in any other.

He goes on to shake the proverbial fist at one of the heroes of post-structuralism. . .

> . . . All that Derrida has done is reinvent the infinite regress. We have no option but to accept the integrity of the original text . . . to deconstruct is not to interpret but to replace with other constructions. . .

Even more derisory, although more humorous and tongue-in-cheek, were the comments of David Myers (1991), a Professor of Comparative Literature at the University of Central Queensland.

> The final goal of our proud deconstructionist revolution is to discover equal meaninglessness and chaos for everybody alike. You can't be more egalitarian and anti-elitist than that, can you?

He continues, mainly writing about post-structuralism's presence in literature, to wax poetically and flamboyantly in a parody of works and ideas he associates with this style of analysis.

> Literary criticism is no longer a fireside chat in leather armchairs at the club. It has become a mother of all wars as power hungry post-struc-

> turalists and spiteful reviewers fight for psychosexual supremacy to the last pubic hair and the last sub-stratum of repressed subconsciousness.

The concern expressed here reflects, to some extent, a conservative reaction to radical critique. The vitriol flows freely between those who believe 'words are what they mean' (essentialism) and those who believe 'words are what we say they mean' (cultural relativism). But another source of this concern derives from the anxiety and genuine dismay of some people toward the post-structuralists' tendency to over-interpret and/or to make claims that those interpretations do more than simply reflect cultural relations, rather they also reproduce them. This claim of reproduction of cultural relations from an interpretation which may not be widely shared, is a substantially controversial and debatable one.

Interpreting the manifest or surface meaning of a photograph or a line of text can be a tricky albeit pedestrian task. But decoding, that act of discerning and describing a meaning beyond and behind the words and images, is truly an abstract and more difficult procedure to follow. Diversity of opinion explodes when theorising about the abstract motives, intentions, schemes or patterns of relations behind our daily objects and social appearances. This way of going about the work of interpreting, this style of decoding and its aids, has led some to charge that these methods are bad ones. But as Deeley (1990, p. 12) wryly remarks 'even bad methods truly reveal'.

Recommended reading

Carney, T.F. (1972), *Content Analysis*, Batsford, London

Deeley, J. (1990), *Basics of Semiotics*, Indiana University Press, Bloomington

Holsti, O.R. (1969), *Content Analysis for the Social Sciences and the Humanities*, Addison-Wesley, Reading, Massachusetts

Lupton, D. (1992), 'Discourse analysis: a new methodology for understanding ideologies of health and illness', *Australian Journal of Public Health*, vol. 16, no. 2, pp. 145–50

Manning, P.K. (1987), *Semiotics and Fieldwork*, Sage, Beverly Hills

Miles, M.B. and Huberman, A.M. (1984), *Qualitative Data Analysis*, Sage, Beverly Hills

Norris, C. (1990), *What's Wrong with Postmodernism*, Harvester Wheatsheaf, NY

Patton, M.Q. (1980), *Qualitative Evaluation Methods*, Sage, Beverly Hills

Strauss, A.L. (1987), *Qualitative Analysis for Social Scientists*, Cambridge University Press, Cambridge

Weber, R.P. (1990), *Basic Content Analysis*, Sage, Newbury Park

4

The written record

Probably the most accessible and readily apparent unobtrusive method of social research is archival and library work. Too often students see the library as a place of textbooks and references for essays and not places of original and often undiscovered data for research. But libraries and archives (for they are not always found together) are places from which scholars from the humanities and social sciences constantly conduct and draw for their original research work. Libraries and archives contain what many researchers refer to as 'existing sources'.

Existing sources include:

- official statistics (government and private)
- books, journals, newspapers and popular magazines
- government, business and other administrative records
- personal diaries, letters and journals

Aside from books and other bound material, records may come in a variety of shapes and sizes including 'drawings, pamphlets, posters, files, charts, volumes, bundles of papers, photographs, invoices, magnetic tapes, films, computer print-outs, microfiche and microfilm' (Pederson, 1987, p. 5). In this chapter I will confine myself to the written record. The next chapter will deal with the audio-visual record such as music, photographs and film.

Ann Pederson (1987) for the Australian Society of Archives

describes four major repositories of information and data in Australia. She describes, among other things, the type of holdings, who can consult these, where one usually studies these documents and who one must negotiate with for their use on a daily basis. A shortened table summarising these areas is reproduced below.

	Registries	*Archives*	*Libraries*	*Museums*
Hold	Active records	Inactive records	Published matter	Objects and artefacts
Access	Employees	Depends on archive policy (usually scholars)	Any member of their community	The public
Availability	May be removed	Search rooms on the premises	Premises or may borrow	Display or exhibition areas
Minders	Records managers	Archivists	Librarians	Curators

(Adapted from a table by Anne-Marie Schwirtlich, 1987, p. 9)

To access the holdings in libraries one uses the library catalogue system and for archives and the like, an index of all the available items might be consulted, if one exists. It is also advisable to have a good yarn to the records manager or archivist for an overview of the holdings and to discuss your research needs and the best or most efficient way to meet these. Obviously for private archives or registries with restricted access, permission must be sought prior to access. For public record agencies (Commonwealth, State or local government archives) an application for daily access is a form-filling exercise, similar to applying for a library card.

Registries and archives are kept by a variety of agencies from Commonwealth, State and local government authorities to universities, churches, schools, private companies, clubs and societies. The Australian Society of Archives recommends that you consult S. Burnstein *et al.* (1992), *Directory of Archives in Australia*, ASA Canberra, for a comprehensive listing.

The advantages of using the above record holdings for your research requirements are several. Holdings can supply both primary and secondary sources of data, plus the literature to analyse that data. Registries and archives can provide basic personal information or descriptive statistics which can then be analysed by you. In that task, the library can supply the academic literature support needed by you to contextualise the study and its findings. Also, some of the data in these holdings may be unique in several ways. The information may be highly personal or it may have been collected in ways ordinarily too difficult or expensive for any one individual to collect. This applies particularly to census data or international data bases (such as the HRAF files which will be discussed below). Furthermore, because archival and museum records are often historical, much of the data held in them cannot be obtained using other methods. Record holdings can also be good sources of longitudinal data, allowing researchers to follow the thoughts of a diarist or the pattern of births, deaths and marriages over time. Despite whatever problems plague the use of existing sources (and there are many as we shall see) such information is highly accountable. Because it is usually fairly available to researchers, other people are able to re-check your analysis and interpretations.

The general advantages of studying record holdings are:

- they provide a comprehensive data source (primary, secondary and support literature)
- the data is often unique (methodologically or historically or administratively)
- they provide a good source of longitudinal data
- they provide a highly reliable data source which can be easily re-checked by others

We will now take a closer look at the three most important types of material held in these various holdings—existing statistics, personal documents and published sources (books, journals etc.) found in libraries.

Existing statistics

C. Wright Mills (1959, p. 205) among others, always argued that you should never do any empirical research unless you absolutely had to. He was always in favour of using available data from official sources and from the published sources of colleagues. He was of the view that we had more than enough data on a whole range of human issues and that the central problem confronting modern social science was that of interpreting it all. 'The purpose of empirical enquiry is to settle disagreements and doubts about facts' he said. Despite this, each year witnesses hundreds and thousands of surveys globally. Each collecting statistical data about hundreds of thousands of people's beliefs, lifestyles, preferences and behaviour. If in doubt, even a casual perusal of *Dissertations Abstracts International* should convince you. The role of budding MAs and PhDs, particularly in the USA, is an overwhelming testimony to the fact that modern humanity is one helluva over-surveyed group.

A chat with any reference librarian will enable you to locate dozens of directories to yet more directories of official statistics and records. The sixth edition of the *Directory of Information Sources in the UK* (Codlin, 1990) for example, lists five and a half thousand organisations which are able to make information available through libraries or publications. Information and records on such diverse topics as AIDS or Bach or little ships are available in works such as these.

Often, official statistics are available by writing direct to certain government departments in Canberra or your State. Alternatively, Australian Bureau of Statistics (ABS) and government census data can be accessed by searching the ABS catalogue in your library and perusing CD–Rom facilities there (for example, Austguide and CDATA). Some of this information and data is sold in government publications and can be easily obtained in their bookshops. Maher and Burke (1991) also provide a more in-depth discussion of these government sources and others such as the National Social Science Data Archives. For less 'officially worked' statistics, some government departments may release current statistics for re-analysing, depending on certain agreed restrictions and qualifications. In this respect, registries of births, deaths and marriages are excellent sources of 'raw' data for analysis, as are wills and probates which can be examined at the Registries of Probate or Public Records Offices around the country. Individual researchers may be at a disadvantage with some of these government departments because the latter sometimes feel that they have greater control over institutions than individuals. This sentiment can lead to these departments preferring to release their data to institutions rather than individuals working alone.

For example, Community Services Victoria (CSV) gave La Trobe University's School of Social Sciences a contract to collect and maintain data describing homeless people in Victoria. Information related to people assisted by CSV was forwarded to La Trobe at monthly intervals and organised into several computer databases.

Because of the sensitive nature of this type of information, it is highly unlikely that a private individual or group of individuals could gain access to this data. La Trobe offered to house and develop the database at very competitive low cost on the provision that its researchers would have first access to the data. The university and its individual researchers are bound to proceed through ethical and bureaucratic committee structures which ensure accountability from the institute and its officers. Under these conditions, CSV was able to obtain the information at low cost, have the data analysed by professionals competent in the area, all the while contributing to the research and teaching goals of higher education and social policy development.

As mentioned earlier, aside from the usual sources of official statistics from government departments, Australia also has a National Social Science Data Archive which is located at the Australian National University in Canberra. This archive produces a guide to its holdings entitled the *SSDA Catalogue* (1991). The catalogue is divided into a number of sections (five in 1991). The first section describes hundreds of studies conducted by governments and universities where these agencies have deposited their actual data files with the archives. Section 2 includes data files of opinion polls conducted by agencies such as 'Morgan Gallup' and *The Age.* Section 3 contains census data and section 4 contains the data files of some *overseas* studies. The final section describes foreign data archives.

To give you an idea of the richness of the studies whose data files are kept by the SSDA look at the following titles (taken at random from the SSDA, 1991, catalogue).

For those who do not have the skills or desire to analyse or

SSDA no.	Title
11	Technical and Further Education (TAFE) Teacher Survey, 1977
12	Renters and first-time home buyers, 1975–76
13	Feasibility of Group Legal Services in Australia: The public's attitude . . . 1980
19	Women in the NSW public service, 1978
27	Canberra Mental Health Survey, 1971
34	Health knowledge and drug use among high school students, 1980
38	Survey of truck drivers on NSW highways, 1976
43	The characteristics of a group of non-drinkers, 1977
111	Vietnamese Refugees in Australia, 1982: The 1st cohort

SSDA no.	Title
115	Images of the Australian Aborigine in illustrated newspapers, 1853–97
152	Australian two party preferred votes
154	Mass media use in Australia, 1976
210	Non-institutionalised aged in Sydney, 1981
211	Images of class in Australia, 1973
226	Blood Pressure Study in Perth, 1975–76
333	Tasmanian Visitor Study, 1981
336	Older people at home in Melbourne and Adelaide, 1981
349	Aboriginal household survey, Sydney, 1984
353	Regular physical exercise by Australians, 1985
375	Australian values survey, 1983
386	Role of Community Pharmacists in Patient Counselling, 1982
405	Social Issues in Australia, 1985
416	Ethno-specific services for the aged in Melbourne and Sydney, 1985
537	Issues in Multicultural Australia, 1988: New arrivals
581	Working holiday makers in Australia, 1986
609	Australian youth survey 1990: Wave 2

re-analyse other peoples' statistics there is also a qualitative–ethnographic version of the above. The Human Relations Area File (HRAF) is a data source/collection of ethnographic descriptions collected by researchers from all over the world. It is divided into regional interests. You will need to check whether your university has HRAF. (For an introduction to this resource see Lagace, 1974.)

There have been a number of problems identified with using existing sources, particularly statistics. The main problem is that methodological errors are inherited and these may be quite large. This also applies to HRAF. The ethnographies in HRAF have been collected with uncertain accuracy and differing purposes (from colonial to anthropological). Much of the material is old and marred by prejudice, ethnocentrism or sexism and must be used with some caution (Zolberg, 1983, p. 69). Martin Bulmer (1984), however, rejects this concern, arguing that the significance of these kinds of anxieties has been highly exaggerated. He points out that errors can be compensated for by re-analysis or by comparative analysis and that these can point to the need for more methodological work of a new kind. Most of the controversy which surrounds the use of official statistics, for example, usually centres around suicide and crime and delinquency statistics—areas where special problems are well known.

The problems of classifications, reliability and validity are not necessarily generalisable to other areas. In any case, even in these areas, generalisations are rarely naive. As Bulmer remarks (1984, p. 186), 'The world is not made up of knowledgeable sceptics and naive hard line positivists'. However, I'm not so sure that Jack Douglas (1967) would be so easily waved away in his deeper concern about the use of statistics in general. Douglas, in his classic review work on Durkheim's *Suicide*, argued that:

> The statistical hypothetical approach fails to take into consideration the fact that social meanings are fundamentally problematic, both for the members of the society and for the scientists attempting to observe, describe and explain their actions. (1967, p. 339)

Douglas argues that ordinary people have constant trouble trying to understand one another and even the day-to-day events of life. How much harder is it, then, for an unobtrusive researcher to comment or explain this from a distance. When not part of their lives, how can outsiders understand. Douglas continues:

> . . . the meanings imputed to suicide independent of concrete situations in which the communicator is involved are different from the meanings imputed to concrete situations in which the communicator *is* involved. In general terms this means that the *situated meanings* are significantly different from the *abstract meanings*.

Statistics—personally gathered or gleaned from existing sources—freeze human action, belief and meaning. They draw our attention from the fluid, processual and ambiguous facts of our lives. How strongly this feature dominates one's presentation or impression depends just as strongly on how dependent one is on numerical data alone.

Some problems with existing statistics are:

- inherited methodological errors
- material may be dated
- material may be collected for different purposes by different agencies
- limited generalisability
- material may decontextualise meaning
- material may artifically categorise and abstract processes which are ambiguous

The use of personal documents may re-inject, re-enliven and bring to life the breath of the real experience in the social and written portrayal. We will now take a closer look at personal documents.

Personal documents

Ken Plummer (1983) provides a nice overview of research which looks at personal documents. To some extent similar to Webb *et al.*'s (1981) work on unobtrusive measures, Plummer's work is a survey piece. He reviews the diversity of 'life documents' and the well-known pieces of research which have used these as the basis of their work. (Plummer also has an annotated 'Suggestions for further reading' at the end of each chapter which is very useful for its specific comments on other work.)

Among the personal written documents one might examine are the life history, the diary, the letter, vox populi (the reproduction of recordings of social life with little or no comment) and oral history. *Life histories* are recorded, in-depth accounts of peoples' lives which have been documented by others. The table below is reproduced from Plummer (1983) and shows a general selection of social science life histories which can be re-analysed for their personal, subjective, or experiential dimensions.

Agnes	A male to female hermaphrodite (Garfinkel, 1967)
Mrs Abel	A woman dying of terminal cancer (Strauss and Glaser, 1977)
Ann	A prostitute (Heyl, 1979)
Herculine Barbin	A nineteenth-century hermaphrodite (Foucault, 1980)
Chic Conwell	A professional thief (Sutherland, 1937)
Cheryl	A young woman in love (Schwartz and Merten, 1980)
Janet Clark	A heroin addict who commits suicide (Hughes, 1961)
Jane Fry	A male to female transsexual (Bogdan, 1974)
Arthur Harding	An East-end underworld figure (Samuel, 1981)
Don Juan	A Yacqui Indian magician (Castaneda, 1968)
Harry King	Another thief (Chambliss, 1972)
Pierre Rivière	A nineteenth-century French family murderer (Foucault, 1978)
Manny	A 'hard core' heroin addict (Rettig *et al.*, 1977)

The Martin Brothers	Five delinquent brothers in Chicago in the 1920s (Shaw, 1938)
The Martinez Family	A poor rural Mexican family (Lewis, 1964)
Sam	A career thief (Jackson, 1972)
The Sanchez Family	A poor urban Mexican family (Lewis, 1970)
Stanley	A Chicago delinquent in the 1920s (Shaw, 1966)
James Sewid	A Kwakiutl Indian (Spradley, 1969)
Sidney	Another Chicago delinquent (a rapist) (Shaw, 1931)
Vincent Swaggi	A professional fencer (Klockars, 1975)
Don Talayesa	A Hopi Indian chief (Simmonds, 1942)
William Tanner	A drunk (Spradley, 1970)
Henry Williamson	A hustler (Keiser, 1965)
Wladek Wisniewski	A Polish emigré to Chicago (Thomas and Znanieski, 1958)

Diaries and other autobiographical sources (letters, journals, for example) may also serve the same purpose by giving us the experiential side of an event, a time or a circumstance. In this category published diaries such as *Diary of Ann Frank* or *Go Ask Alice* should not be overlooked. Suicide diaries and letters which are reproduced in part in journals such as *Death Studies* (USA) or *Omega* (USA) are also useful. There is also a growing body of self-confessional literature on the social science 'field' experience which can be usefully analysed (see, for example, Bell and Encell, 1978, and Perry, 1989).

Letters are also useful sources of unobtrusive data but are much less used in social sciences than diaries. Almost everyone who writes about document analysis in the social sciences refers to Thomas and Zuaniecki (1958) *Polish Peasant.* This was a study of hundreds of letters between Poles and Polish emigrés to the USA.

Vox populi material is quite popular now although Studs Terkel is known for pioneering this type of work. Running all over America and jabbing a tape recorder microphone in peoples' faces before darting toward someone else was a favourite method used by this flamboyant social researcher. Popular feminist material has also produced quite a few books which simply reproduce interviews with women 'who chose not to have children' or people who have 'recovered from nervous breakdowns' and so on.

Oral history can be found regularly in the *International Journal of Oral History* but there is also an increasing volume of books and recordings devoted to this interest. All of these can be analysed and re-analysed for their lived-in experience content.

If the material (the personal documents) have been published then the task is merely to track down and locate the items through normal library and inter-library channels. If unpublished then the task is probably an archival one. John Stanfield (1987), in an article in the *American Behavioral Scientist*, pp. 369–72, provides advice if you are heading toward archival work to find your letters, diaries and assorted odds and ends.

Stanfield's advice about archival work is:

- review the secondary literature on the topic first
- take notes on footnotes and bibliography for clues to archival sources and retrieval clues
- take note of the culture, demography and time-specific features of the period, place and person

According to Stanfield, this advice allows you to: (1) develop a preconceived but informed theoretical hunch/or conceptual framework; (2) to establish an informational bench mark to assess archival material ignored, devalued or exaggerated; (3) alert you to gaps in the current state of knowledge on a particular topic.

Following the review of secondary literature:

- spend the first day or two getting to know the archivist
- familiarise yourself with the register/index of the archive holdings

Preliminary discussions with archivists can save valuable time as they are, after all, the best informed minders of the material. They can provide valuable information and gossip about a research issue which can be very helpful indeed. Careful study of the register or indexing system can also be valuable 'not only to pinpoint the obvious' but also to bring 'to mind the importance of the less obvious' (1987, p. 372).

Of course, as with the use of official statistics there are a number of problems. Among the most prominent hitch is the issue of bias. Autobiographical material such as diaries and letters are, of course, usually confined to the literate. Of the literate, diaries predominate among skilled manual workers and upper-class domestic servants.

Historically, women have been less well represented (Scott, 1990, p. 178). Letters have mainly been the collected province of the upper classes until (quite recently, that is), this century. Despite this problem, Webb *et al.* (1966) remind us to keep our nerve—'bias itself is not fatal' they remark 'only not knowing it is' (p. 108). And despite the fact that selectivity operates through others, written records are probably no worse than interviews that rely on the memory of the respondent (Webb *et al.*, 1966, p. 111).

There are, of course, other more practical problems which are somewhat unique to this kind of work. I remember when I spent some time in the archives of the Mitchell Library in NSW looking for nineteenth century evidence of beliefs about dying, I located a nineteenth century diary that the register suggested might have comments about death and dying inside it. I lodged a retrieval request. I waited for nearly an hour. When I finally got hold of this wonderful document, I couldn't decipher the handwriting of the diarist. It may as well have been in ancient Greek for all the good it would do me!

Even when reading the handwritten material is technically possible, the issue of literal meaning may still be a problem. This can be exacerbated by the author's use of codes, personal shorthand, abbreviations or simply the taken-for-granted meanings between writer and recipient (Scott, 1990, pp. 179–80).

Finally, apart from the unconscious or unintended biases within letters, diaries and autobiographies there is the problem of intended, conscious attempts at impression management. Letters describing an event or feelings are not the same between author and mother and author and lover or distant friend or business associate.

Stanfield warns that the only real way one can dissipate or guard against these and other effects is to *combine* your methods.

The advice against over-reliance on one method or one source of data is a good one—for all social science methodology. For archival validity, remember that 90–95 per cent of records (particularly institutional) are destroyed (Pederson, 1987, p. 35) as a matter of course. Archives contain only a fraction of the surviving material and much remains in personal/private hands with highly protective families who attach great sentimental value to this material.

Again, unlike the use of official statistics, there are additional practical problems with archival work. I spoke to historian Joan Beaumont, the author of *Gull Force* (the events upon which the film *Blood Oath* was also based) about her archival work as a historian. For *Gull Force*, a study of Australian POWs in World War

Methodological problems associated with archival work are:

- authenticity of the documents (see Brooks, 1969)
- representativeness and bias
- understanding the literal/intended meaning (emic)
- reading/legibility problems
- incomplete sources

To guard against these problems:

- be familiar with the secondary literature
- cross-check your findings with interviews or other contemporary accounts from similar persons
- combine archival work with other methods if possible/feasible

II held by the Japanese, Joan studied private diaries from camp commanders and ordinary soldiers. She also examined the official reports by officers concerning illness, rations and atrocities experienced by the POWs. These were made for the purposes of the War Crimes Trials. Much of her work was conducted at the Australian War Memorial Museum. In this work and other work throughout her career, there have been several practical annoyances, first unanticipated, and later anticipated, but time-consuming nevertheless.

The newcomer to archival work should be aware that:

- Not all material is available to the public. Some of the material may be censored by governments because of the sensitive political nature of the content. Other material may be withheld by families for their own personal reasons. Permission can often be difficult to obtain.
- Some censored documents may be released but need vetting by the authorities to determine this. That process can take months.
- Even when access is available you may need to obtain copyright permission from the owner/s of the diary or letters to reproduce the material. And by 'reproduce' I mean actions as simple as photocopying.
- Exacerbating the previous problem is that the copyright owner may not be known.
- Archival work produces volumes of material and this creates practical problems of filing, retrieval and repetition.
- The volume/size of the material also means that photocopy

Archives contain only a fraction of the surviving material and much remains in personal/private hands. Courtesy of the *Geelong Advertiser*

budgets may be considerable. Some of this may be offset by reading some of the material onto tapes or typing it in directly to a lap-top computer.

- Because archives are dispersed all around a country, travel budgets can easily blow out.
- The contemporary practice of placing many documents (for

example, newspapers, minutes of meetings) on microfilm means: (1) harder work on the eyes; (2) headaches; (3) difficulties in comparing two documents or more side by side. Watching moving print all day can be quite nauseating.
- Archival work can also be extremely time consuming. Ordering from the index/register, requisitioning and delivery of the documents can take hours—and then the documents may not be what you thought or needed.
- Compounding the retrieval/time problem is the fact that some material 'held' by the archives may not be actually 'located' there. The actual documents may be housed in other buildings and may take a day or so to retrieve.
- Archives keep restricted hours, many not opening at night for example. Stocktakes may also close or restrict access even further. Many archives are in administrative areas of cities with few shops. Take your own lunch and Minties! And remember, no bags allowed in search rooms.
- Finally, get used to using pencil. Many archives don't allow you to use Biro or ink pens for fear you might go berserk with them on one of their antiquarian manuscripts.

Given the multiple problems involved in archival work, particularly in accessing personal documents, is it really worth doing this kind of work? Ken Plummer (1983) argues that there are at least five advantages in preserving a search and analysis of personal documents for their qualitative input to research. (These are listed on page 65.)

Let us look at a few examples of how personal narrative is able to perform some of these functions. C. Wright Mills (1951) in his classic study of the American middle class peppers his theoretical and politico-social analysis with vignettes of the people he is analysing. In one section he draws on the published descriptions of saleswomen aged between 18–30 based on the observational work of J.B. Gale (pp. 174–5).

> THE WOLF prowls about and pounces upon potential customers: 'I go for the customer . . . Why should I wait for them to come to me when I can step out in the aisles and grab them? The customer seems to like it; it gives them a feeling of importance. I like it; it keeps me on my toes, builds up my salesbook . . . the buyer likes it too . . . every well-dressed customer, cranky or not, looks like a 5 dollar bill to me'.
>
> Intensified THE WOLF becomes THE ELBOWER, who is bent upon monopolising all the customers. While attending to one, she answers

To search for the personal, individual account helps to:

- rehumanise a portrayal of experience by going beyond images of mass, periodicity and institutional changes and statistics.
- explore issues which are not preconceived or if preconceived may stretch beyond the expected. This may offer or stimulate further theoretical and conceptual processes.
- complement and balance portrayals of the many with the few, the personal with the impersonal, the social abstract reality with the personal lived-in one.
- clarify and consolidate the understanding of human experience through the technical use of empathy and reader identification.
- enliven and enrich any social science discourse making such discourse engaging and compelling, whether for research or for the purposes of teaching.

> the questions of a second, urges a third to be patient and beckons a fourth from a distance . . . THE CHARMER focuses the customer less upon her stock of goods than upon herself. She attracts the customer with modulated voice, artful attire, and stance. 'It's really marvellous what you can do in this world with a streamlined torso and a brilliant smile. People do things for me, especially men . . . after all a girl should capitalise on what she has, shouldn't she?'

Page after page Mills takes us through each character on the salesfloor—the shy salesgirl, the university student part-timer, the floor drifter, the social pretender, the old timer—so that in several pages an image beyond the simple one of organisation is created—the culture of floor sales.

Historian Dee Brown writes 'an Indian History of the American West' (1975) but his descriptions of tragedy and crime against the Indians are so vivid that outrage and grief appropriately accompany intellectual understanding by the reader. A report by a white soldier records part of a massacre:

> The Navahos, squaws and children ran in all directions and were shot and bayoneted. I succeeded in forming about 20 men . . . I then marched out to the east side of the post; there I saw a soldier murdering two little children and a woman. I halloed immediately to the soldier to stop. He looked up, but did not obey my order. I ran up as quick as I could, but could not get there soon enough to prevent

> him from killing the two innocent children and wounding severely the squaw . . . (p. 16)

Many Indians 'realised' that something momentous and tragic was happening to their nation, but to read some personal reflections of some of their leaders rehumanises the meaning of words such as 'recall' and realise. This reflection by Black Elk quoted in Dee Brown's history is a good illustration (1975, p. 353).

> I did not know then how much was ended. When I look back now from this high hill of my old age, I can still see the butchered women and children lying heaped and scattered all along the crooked gulch as plain as when I saw them with eyes still young. And I can see that something else died then in the bloody mud, and was buried in the blizzard. A people's dream died there. It was a beautiful dream . . . the nation's hoop is broken and scattered. There is no centre any longer, and the sacred tree is dead.

It is difficult to imagine a social science language which could match the prose and impact of Black Elk's description of the tragedy which is his people's recent history.

In an entirely different context, Sam Keen reminisces about childhood and attempts to convey the sense of magic and serendipity which is part of that time. How to convey that magic and surprise? He recounts a story from when he was just six years old.

> When I was six years old I was walking by a courthouse in a small town in Tennessee. A man came out, followed by a large crowd. As he walked past me, he pulled a knife from his belt and said, 'I present you with this knife'. Before I could see his face or overcome my shock and thank him, he turned and disappeared. The knife was a strange and mysterious gift. The handle was made out of the foot of a deer, and on the blade there was something written in a foreign language which no one in town could translate. For weeks after this event I lived with a pervasive sense of gratitude to the stranger and with a wondering expectancy created by the realisation that such a strange and wonderful happening could occur in the ordinary world of Maryville. If nameless strangers could give such gifts, what surprises might be expected in the world? (Quoted in Savary and O'Connor, 1973, p. 259)

Finally, Louise Zaetta (1980) collected reminiscences from adult Catholics about their childhood to illustrate her exploration and satire of Catholic upbringing in the 1950s and 1960s. The stories children swapped with each other to ensure that each conformed to Catholic teaching are instructive because, Catholic or not, we can all identify with similar foolishness and fears of growing up and developing a basic and workable morality.

> There was a girl who was told not to play tennis on Good Friday. Unfortunately, she disobeyed. Just as she was about to serve, she was turned to stone, and they had to squeeze her into her coffin with her arm raised, in the serving position, her racquet still in her hand. (p. 42)

> Wherever you died, even if it was in the middle of the Sahara Desert, if you [are] wearing your Scapula, a Catholic Priest would be there to hear your last confession. (p. 43)

> If you buy a coloured Holy Picture (card) instead of a black and white one, you'll be assured of gaining a higher place in Heaven. (p. 44)

About the only problem with the use of personal documents, used to advantage in the above way, is that illustration can sometimes stand in for analysis—and in social science this is not desirable. The task of the human sciences is to see the many in the few and to connect one experience to that of other lives, showing what is unique alongside that which is common. Furthermore, use of personal documents/narrative can expand the volume of any analytic work and this can create problems for journals whose bias is quantitative or numerical. In health and medical journals, for example, this severely limits qualitative research presentation, including those which use document analysis (Daly, 1993).

Published sources

If using existing government and other data archive statistics is not your cuppa, and in any case archival work seems a lot of trouble, much good work can still be conducted in the main part of the library. If the books and journals in the library are seen as data rather than as simply authoritative references, the library takes on new meanings. Let me provide some examples of what I mean.

Social worker Jan Fook has spent a substantial part of her career advocating the need for social casework in social work practice to move away from the psychological, counselling model it seems to be currently embracing. She has argued that there is a need for work with individuals to be political and sociologically informed and empowering for clients (Fook, 1993). In one of her papers which appeared in the journal *Australian Social Work* (Fook, 1991), Jan decided to see whether or not university and CAE courses in social casework had learnt anything from the radical and feminist critique of the 1980s. Is casework becoming more 'responsive to the current social and ideological context'? (Fook, 1991, p. 19).

To answer that question Jan collected the subject outlines of social casework for all the major social work and welfare work courses in Australia. Some of these were obtained by asking the institutions concerned to send these to her but when these were not forthcoming she simply looked them up in the *university/college handbooks* in the library. By studying course outlines and subject descriptions for her casework interest, she was able to discover how casework is currently defined and its relative importance in the structure and assessment of a social work/welfare degree. This then formed the basis of further criticism for her in this area.

Sociologist Glenda Koutroulis was interested in 'sexist ideology' in medical practice. To pursue this interest Glenda identified the main *textbooks* used by Australian medical students when studying obstetrics and gynaecology (Koutroulis, 1990). Following the semiotic tradition of analysis, Glenda's aim was to 'produce a reading for a hidden curriculum of sexist ideology' (p. 74). And find that 'curriculum of sexist ideology' she did! The texts were laden with sexist comparisons between women and animals, stereotypical remarks and gratuitous references to women as identical to wives, mothers and homemakers. Few of the texts mention or integrate the insights of Kinsey or Masters and Johnson let alone feminist writers.

Educationalist Derek Colquhoun examined healthism in the physical education curriculum (Colquhoun, 1990). Healthism is the moral sentiment attached to being fit and healthy and the formal pressures and expectations brought to bear on deviants such as fat people, smokers and happily sedentary people. Derek studied the *Body Owners Manual*, a book used in physical education classes for young Australian children. This key textbook is illustrated by cartoons and uses the motor car as a key analogue for the body. In semiotic tradition, Derek looks for the 'hidden' curriculum of healthism, the moral baggage which comes with the funny cartoons and clever lines and exhortations to keep fit and have a 'good' body shape.

Looking for how interested the Commonwealth government was in rural health, I examined four national *policy documents* on health. These included the 'Health 2000: Health for All' document, 'National Mental Health Policy', the 'National Womens' Health Policy' and the 'Rural Book' of health services. My task was a simple content analysis (Kellehear, 1990). Who mentioned rural people and what did they say? I found that 'Health for All' was not for all, it omitted some people, particularly rural dwellers. The 'Rural Book', with a few minor exceptions, boasted about services available to everyone rather than mentioning region-specific assistance. The other policies had idiosyncratic and conflicting definitions of the word rural which allowed them to make strange classifications. Wagga Wagga (pop. 50 000) was classified in one document as urban whereas Newcastle (pop. 250 000) was classified as rural. This formed the basis of my published critique of what I saw as complacency and apathy.

Textbooks, policy documents or tertiary institution handbooks are not the only sources of data. Anyone in the humanities will tell you that *fiction* is also a wonderful source of cultural analysis and data. Shirley Walker's (1983) *Who is She? Images of Women in Australian Fiction* is a good example of several authors using fiction as the basis of their social analysis. I reproduce on page 70 the contents for you to browse because this one example gives you fifteen others to think about.

Newspapers are also a rich resource for unobtrusive research. Continuing the example of researching women, Kate Smith (1981) studied the media's attitude to women. This project involved a content and thematic analysis of newspapers in Adelaide over a period of fifteen months. And, of course, on another topic, the National Social Science Data archives has research reporting on images of Aborigines in newspapers spanning the late 19th century (SSDA Catalogue, 1991, item 115). I'm sure most of us can think of many more examples of research which have as their data source fiction, newspapers, popular magazines and so on.

In summary then, archives and libraries may be sources of data in terms of primary sources (ethnographic–statistical, diaries and letters) or using so-called secondary sources as primary data sources (newspapers, textbooks, novels etc.). In this way, registries, archives, libraries and museums may supply both the data and the means to analyse it, all within the close proximity of library and archive buildings in one or several cities. That's about as close as one gets to one stop shopping in social research terms. But written documents are not all one gets from these sources. In the next chapter, we will look at audio-visual sources of data.

Ethics in the library

Ethics is a difficult subject. But as I argued earlier, the social and

(From *Who is She? Images of Women in Australian Fiction* by Shirley Walker, 1983.)

moral implications of all research should be discussed with friends and colleagues. The ethical issues that you might consider in research with existing sources are similar in category, although different in detail, to all social research. *Cheating* is a concern. One must be careful not to misrepresent data, particularly data that is obscure in location, that is, not readily available. Hence the researcher's obligation and commitment to honesty is in the spotlight. In the same vein, one should not ignore copyright restrictions. However cumbersome those restrictions might seem they do protect the privacy and ownership of others. Every reasonable attempt to ascertain the copyright owner should be taken and permission then sought. Finally, one should take care not to *plagiarise* the work of others. Ideas and tracts of text from the written records of libraries should always be acknowledged with the appropriate referencing in text.

Consent remains an ethical concern, not simply in copyright terms, but also in the avoidance of misrepresenting your role and purpose to officials in libraries, archives and registers. In this respect, a dim view is to be taken of bribery. *Confidentiality* remains a relevant ethical consideration especially around concerns about whether those named in the documents wish to be identified. Indeed, some owners of documents sometimes do not wish to be identified for security or publicity reasons. Fook (1991) for example, in her study of university courses described in handbook guides, did not identify the institutions which conducted those courses. Negative conclusions may lead to poor publicity, lower enrolments and consequently to laying off of contract staff.

Do not think that simply because you are dealing with books or statistics that people are not affected by your presence or by what you subsequently write. Always attempt to think and discuss the issues with others. Begin with the above ideas and place them in the context of your own individual project. What other ethical issues, not identified by me, might impinge on work with existing sources?

Recommended reading

Brooks, P.C. (1969), *Research in Archives*, University of Chicago, Chicago

Dale, A., Arber, S. and Proctor, M. (1988), *Doing Secondary Analysis*, Allen & Unwin, London

Maher, C. and Burke, T. (1991), *Informed Decision Making*, Longman Cheshire, Melbourne (see chapters 1–3)

Plummer, K. (1983), *Documents of Life*, George Allen & Unwin, London. (Has an excellent bibliography and annotated suggested readings at the end of chapters.)

Social Science Data Archives, *Catalogue S.S.D.A.* (1991), the Australian National University, Canberra

Stanfield, J.H. (1987), 'Archival methods in race relations research', *American Behavioural Scientist*, vol. 30, no. 4, pp. 366–80

Stewart, D. (1984), *Secondary Research: Information Sources and Methods*, Sage, Beverley Hills

Burnstein, S. *et al.* (1992), *Directory of Archives in Australia*, ASA, Canberra

Wright Mills, C. (1959), *The Sociological Imagination*, Oxford University Press, New York (see especially, appendix)

5

The audio-visual record

Libraries and archives, and personal and public records, are not simply filled with words and numbers. They also contain sights and sounds. The visual image and sounds, particularly music, are part of the audio-visual record of human culture which is too often missing in contemporary social science research. Rob Walker (1991), a professor of education at Deakin university, sums up this situation perfectly:

> The visual has become silent in the social sciences at the very point in history when it dominates both science and culture . . . curious.

The neglect of the visual in social research is ironic given that social sciences have a tradition of aping the physical sciences (Albrecht, 1985). Botany, archaeology, molecular physics, engineering, sports physiology and astronomy are only some of the sciences that place the visual at the centre of their methodological concerns.

Furthermore, similar to an examination of the written record, investigation of the audio-visual record has several distinct advantages for the unobtrusive researcher. First, visual sources such as film and photographs are available for re-analysis and re-checking by others. Second, images allow researchers to discover previously unnoticed or ignored aspects of a scene or portrayal. Inter-rater tasks (using other people to offer their opinions about what they see) are easily performed to enable the overlooked to emerge even when one researcher repeatedly misses certain detail. Third, the study of images or music can be important clues to the cultural world of people whose communication emphasis is not strongly placed on speaking, for example young children. This highlights a broader strength of using these sources, that of drawing cultural insights from non-literate forms of communication. This is a particularly important issue when analysing non-academic, non-professional cultures. Popular culture is largely audio-visual. It employs television, cinema, videos, photographic advertising, music, sound effects and live theatre to convey important cultural messages. Examining some of this record directly is treating this audio-visual experience as primary data rather than analysing it in a secondary way such as depending only on interview accounts of it from other people. Audio-visual sources can also be used to cross-check, support or challenge information gained from written records.

Audio-visual records as a source of unobtrusive research:

- can be easily re-checked and re-analysed by others
- allow greater detail in observation and analysis to emerge
- are important non-literate source of information about people, especially less literate and verbal groups .
- allow researchers to assess wider cultural themes in broader ways, i.e. through popular sumbols and icons in film, television and music
- can be used as alternative and complementary sources of information to written sources

In this chapter, I will concentrate on three examples of the audio-visual record to further illustrate how these sources can be valuable for the purposes of social analysis. I will discuss photographs, film and television, and music as sources of cultural knowledge for the unobtrusive researcher. In each section, I will mention some selected studies that have employed these data sources in one way or another. Sometimes those researchers have employed content analysis or simple thematic analysis. At other times, particularly in examination of film and television, the semiotic approach seems dominant. The unique problems associated with analysing these particular sources of data will be discussed. Towards the end of each section I will list some basic suggestions or steps on how to begin a preliminary analysis of your own.

Photographs—a still image

The still image abounds in most human societies and is not, of course, confined to the recent innovation of the photograph. The sociology of art (Zolberg, 1983) is a field which, like many other areas in the humanities, is based on the study of the image—notably painting, drawing and sculpture. Covey's (1991) recent study of images of old people in Western art is a good example of a multidisciplinary approach to the visual. Webb *et al.* (1966) also suggested the study of art work in their first edition and they extended this idea to children's drawings. They cited a study which examined children's drawings of Santa Claus both before and after Christmas. Contrary to my own waistline experience, children tended to draw Santa larger before Christmas and smaller afterwards. The history of interest in children's drawings for what they may tell us about the social and personal life of children dates back to the nineteenth century. For a review of some of these issues, readers might like to look at the recent work of Thomas and Silk (1990).

And, finally, the importance of photographs as valuable sources of cultural analysis is recognised in the recent move by the Australian National and State Libraries to establish a video disc service. By visiting your State library (for example, Mitchell Library in Sydney or La Trobe Library in Melbourne) one can access this service by merely asking. By operating the compact disc terminal, upwards of 100 000 still photographs can be viewed and searched in a fly-by-slide fashion. The computer searches for images based on your keyword input (for example, 'rural', 'women', 'immigrants'). The photographs can then be 'thermal' printed on fax paper for a rough copy reference.

The sociologist Erving Goffman (1979), apart from his famous studies of stigma and asylum behaviour was also well known for his study of another type of image, that of advertisements. He analysed these for depictions of gender. Recently, Wernick (1991) has continued this tradition. But away from the more direct concern for media and art studies, photographs have received very little attention—little, that is, compared to the enormous and widespread experience of its practice among ordinary people. Since George Eastman invented the box brownie back in 1888, the photographic image has permeated every aspect of our lives from passport, student and drivers' identification cards to family and travel home albums. Social scientists have done little to tap this.

Two important studies to emerge recently illustrate the potential of the photograph as data sources for cultural studies. Dowdall and Golden (1989) used over 300 photographs from a pool of 800 to analyse the early history of an American mental hospital. They were interested in the hospital's social and ideological features. The photos were collected from the hospital office of public information, from staff donations, from printed sources such as medical journals and annual reports, from major photographic collections in the local community and from antiquarian book sellers. They compared these images with the written documents of the times—from newspaper accounts, hospital reports, patients records and staff publicity.

Employing basically a semiotic analysis, Dowdall and Golden noted the sex ratios, the staff patient ratios, urban rural content, and personnel and activity content of the images. They found the sex ratios to be misrepresented in the photographs, with more men than women appearing; that the hospital projected a rural image when, in fact, it was situated in a city; that doctors were a rare find in a photograph and that the main image was one of attendants and patients. This conveyed a much more custodial and coercive image than the written one of therapy and medicine. The images helped to deconstruct the public relations image portrayed in the written materials of the time.

Technically more thematic in approach, Walker and Moulton (1989) examine personal photograph albums. They argue that such albums function to preserve the moment, the feeling and the memory—they are sentimental constructions. Collections tend to signify more than individual images and, like Walker (1991), these authors emphasise the importance of using a narrator, preferably the owner of the album. Williamson (1986) from the semiotic tradition, is not so sure about the importance of the narrator.

Families can be difficult places to be. The real question for Williamson (1986) is *whose* photographs are they? Who is taking the shot? Family albums particularly represent the hegemony of one class—commonly parents or schools. What is being hidden by the sentimentalised construction and portrayal of the family album? Where are the angry parents, the crying or teeth-gnashing children, the uncle with wandering hands? But she concedes that sometimes personal photographs can reveal as well as hide. Verbal myths can be exploded by photographic evidence. Williamson offers a personal account of her families received wisdom of how well she accepted her newly born sister. Yet every image—not in the family album—reveals her jealous, anxious, suspicious and sulky looks toward the newborn!

In the same work, Williamson (1986) goes on to analyse images of women in the photography of Cindy Sherman. This is a clever and clear piece. In a twist to the deconstructive approach, Cindy Sherman constructs—creates—images of women from various contrived self-portraits. In that photographic essay, she shows how the images are artificial constructions set up to receive other people's projections. 'Your reading is the picture' argues Williamson (1986, p. 95), 'the stereotypes and assumptions necessary to "get" each picture are found in our own heads'.

Of course, not all pictures are, or need to be, of people. Schwartz and Jacobs (1979) detail a photographic essay of signs of prohibition—signs which say 'no' round the city. They display signs such as 'no parking' and 'no entry' with others such as 'no hawkers' and 'no junk mail'. If, as Durkheim thought, increasing progress is associated with greater equity and liberality then, according to Schwartz and Jacobs, we're in a lot of trouble.

Walker (1991) argues that there are basically two types of photos, the 'look at that' image of the commercial photographer and the 'remember when' image of the personal amateur. The former are common in artistic, media, advertising or professional photography where information content is primary. Many of the mental asylum photographs were of this type. Personal photos, on the other hand, do not place as much importance on technical or aesthetic quality—it is not the details that matter as much as the context which is remembered. The finger shadow over the snap-shot image contributes rather than detracts from memory.

Walker (1989) like Sontag (1979) before him argues that photographs are evaluations, they are personal, subjective, moral narratives. They tell a story, or rather they help to tell a story which is highly subjective. Photographs do not simply record because no two images are the same. Photographs document the sight and the seer, the style of visualising as well as the visualised.

Giving people a large number of photographs, or asking them to tell you about their photograph albums, tells you about their categories of experience, belief and feeling, their priorities and values. Walker regards photographs then, as the principal 'instruments for the recovery of meaning'. However, Schwartz and Jacob (1979, p. 86) warn that when the interpretation is interpreted it is not always easy to distinguish between researcher and researched narratives. This highlights other problems with the use of photographs as data sources.

Scott (1990) spends quite a bit of time being anxious about the problem of representativeness. Not only is a photograph a subjective and highly personalised example of subjectivity, this selectivity does

Of course, not all pictures are or need to be of people. Courtesy of the *Geelong Advertiser*

not stop at choice of subject matter. Beyond props, choice of subject and setting, other issues are critical in the impression given to others by photographs. Issues such as perspective, lighting, focus and depth of field, the portrayal or omission of movement or context, and the tampering with colour and effect through the use of lens filters are among only the few one might mention. In respect of media use of photographs, selectivity can also operate at the level of editorial bias in choice. Webb and his colleagues (1981, p. 123) discuss several examples of this in newspaper photographs of Edward Kennedy and Malcolm X. And all this before discussing even the basics of darkroom magic and trickery. The photograph is very much a social construct. As Scott (1990) phlegmatically opines, the camera may not lie but who knows about the photographer? Remember, the shot of Uncle Bob drinking a Coca-Cola with Michael Jackson? Remember Suzy strolling along the deck of the superliner QE2 (before they cleared the deck of visitors)? Remember the shot of your mate Gary behind the wheel of a red Porche (before the owner grabbed him by the scruff of the neck)? Photograph albums may also be weeded and reshaped and reorganised when self-image, sentiment and friendships and associations change. Photographs are not simply records, they are play things, sentimental things, social and often deliberate acts between consenting friends and family.

On the other hand, Dowdall and Golden (1989, p. 186) argue that representativeness is beside the point. As long as photographs extend and deepen our understanding beyond the observable or the written record, the effect is valid and useful. Furthermore, as Williamson (1986) has shown in her own account of sibling rivalry, home photographs may reveal—representativeness be damned. Home photographs often are taken during rites of passage (birthdays, weddings, holidays, Christmas etc.) and although albums are contrived they are also far from technically perfect. This lack of technical and formal control over the images themselves and their placement in albums means that often conflicting evidence is available in parts of a picture, or an album, or away from the album but somewhere else in the house. Even some formally posed for, commercially taken photograph of the nineteenth century will sometimes reveal the shabby and ill-fitting clothes of those in very old 'Sunday best clothes'. These shots can reveal the aspirations rather than the actual class positions of those people (Scott, 1990).

Photographs can also be a problem in another cultural sense. Not all cultures easily read or understand photographic images.

Schwartz and Jacobs (1979, p. 87) cite a study of Africans shown a demonstration film on how to rid their houses of standing water. After viewing the film, the audience confessed to not being able to see the procedure. They did however all see a chicken. It emerges that these people read the film in part segments across the image—much like a printed book. They did not take the film images in as a whole. The screen was not viewed totally but rather scanned for small detail.

Scott (1990) cites another example from Cameroon. The Dowayo people could not distinguish men or women or animals from photographs or drawings, despite managing to do so in reality. When identification cards were issued by the government, people often interchanged these as the officials responsible for checking them were also unable to tell whether the photograph bore any resemblance to the carrier.

These are good, if extreme, examples of a limitation which can be used to advantage. The point is that not everyone reads a photograph in the same way. 'Seeing' images is a social/cultural practice which is learnt and is learnt differently by different people (Bourdieu, 1990). In pursuing a semiotic analysis, for example, one can use outsiders to see things that we might not notice. For example, the organisers of the demonstration film did not notice any chicken until this had been singled out. The use of outsiders (of family, of culture) can supply information about taken-for-granted images which can act as 'counter factuals'. The counter-factual is an object, relation or perspective which challenges the prevailing meaning of those who claim to 'know'.

Problems in analysing photos are:

- representativeness—how generalisable, selective, hegemonic the portrayals are
- authenticity—how genuine the associations or images being projected are
- emic/etic (especially for historical period shots)
- emic/etic (for those socially outside the group)

Finally, when looking at photographs, how do you begin to analyse these as data sources? The list on page 82 is a selection of practical suggestions drawn from Plummer (1983) (quoting from Akeret, 1973), Curry and Clarke (1983) and Walker and Moulton

To help overcome problems in analysing photos:

- know your subject (the time period, the culture, the fashions, the problems of the day, attempt to close the gap between photo and context with homework about the social type or circumstance), (Zolberg, 1983).
- avoid being over-concerned with representativeness and attempt to understand the categories used to organise an exhibit, an album, a collection or, in the case of a single image, the 'composition'.
- for outsiders' photographs use history and informants (if alive) to discover emic. Use your etic to reveal the hidden or repressed or deliberately suppressed.
- for insiders' photos (yours, personal or cultural) use outsiders for counter evidence and perspective. Use your own feelings and thoughts for the emic.

(1989). Their suggestions overlap a bit and I've left some out and developed others and collapsed yet others together. You should go to the original references for more detail and, of course, these suggestions should accompany the techniques of pattern recognition discussed in chapter three.

This list, as are those mentioned by different authors, is simply a preliminary one. You may add others, see for example Collier and Collier (1986). Graeme Burton (1990, pp. 12–13) provides an excellent example of how to analyse a photograph. His example uses an advertising image and his style of analysis is worth studying. Whatever approach you adopt the main task is always to keep before you the aim of linking personal or isolated meanings and images to broader cultural or other personal orbits of social life.

Film and television—the moving image

Most libraries have a television or two and nearly all of them have videos and films of one sort or another. Libraries regularly order from film dealers such as the Australian Film Institute. Hiring fees vary from $50 to $60 to several hundred dollars for institutions. However, the lower prices are for public libraries, schools, TAFES, private groups or individuals. Film is usually available on VHS video and 16 or 35 mm film reels. These are particularly useful if the library pays for these or if you are interested in films which are not

Practical principles in analysing photographs

- Decide what type of collection it is (i.e. commercial, advertising, family album, travel album).
- What is obvious? What is subtle?
- Note body language, setting.
- Note themes of change, continuity, outsider/insider similarity and difference.
- Note prominence and timing of status symbols and possessions, technologies or other unusual objects.
- Note repetition of locales, people, objects, lifestyles.
- Note the perspective—close-up or wide-angle.
- Note what attracts you, disturbs you, moves you.
- Note what reference groups are suggested by the furnishings, places, clothes and accessories of the people.
- What type of gender and age relations are suggested.
- For travel photographs, note scenery or 'life away from home' type shots.
- Note your own reactions carefully and reflect about what personal or social values are stirring these.
- What social story or message is intended? Is there another social story or message that can be told? What is this story?

commercially available—some documentaries, government or training films. Do not forget also the local humble video 'library' which is a good, often cheap source of commercial film.

In addition to these sources, Australia has a National Film and Sound Archive in Canberra. That's the good news. The bad news is that collecting and lending by this archive has slowed almost to a stop in the last couple of years. This is because they have been busy restoring and copying historical film and sound footage before they disintegrate. Nitrate film, lacquer discs and wax phonograph cylinders are all fast corrupting and moulding. Their duplication is a race against time (Haigh, 1992).

Viewers of the Italian film *Cinema Paradiso* will remember how volatile and inflammable were those old nitrate films. So much so that, in Australia, these strips of nitrate were used to ignite and sustain fire scenes for other film sets. The incendiary role of our old film has ensured that only about 5 per cent of Australia's early cinema efforts have survived (Haigh, 1992). For sound recordings, this figure may be even lower. For example, only twenty of the

6000 episodes of Australia's longest radio saga 'Blue Hills' have survived.

The archive collects film from all traditions of Australian film and has even recently added to its collection a silent pornographic film. However, since this is not dated, credited or titled, the archives are not absolutely certain of its authenticity or Australianness. Ray Edmondson, the deputy director of the archive, talked to *The Age* reporter Gideon Haigh about that film—a film regularly ordered by those claiming to be 'historians'.

> 'We're pretty sure it's local though. It would seem she said. 'The girls wear socks to protect themselves from bindies.' (p. 7)

Television and film criticism is a huge area and has many traditions within it: narrative theory, reader-oriented criticism, genre study, ideological analysis, psychoanalysis and, of course, feminist criticism (Allen, 1987). For a discussion which links this theoretical diversity with differences in methodological approaches you might consult some of the introductory texts in the area such as Dyer (1982), Eagleton (1983) or Fiske (1987). Most of these are either informed by or begin with a semiotic approach. A very simple and clear introduction to how semiotics works around a television example is provided by Seiter (1987). Following Christian Metz (1974), Seiter focuses her analysis on image, graphics, language, voice, music and sound effects. She analyses the opening one minute of the credits in *The Cosby Show* and demonstrates how certain elements of image, music and lighting convey the Cosbys as wealthy individuals in a utopian family.

> Harmony is the rule; play abounds as a means of solving discipline problems; marriage is sexy; gender equality is the stated goal; parents and children enjoy stimulating and satisfying situations at work and in school; childcare and housework are either invisible or enjoyable. (p. 38)

Mayer and Burton (1991) also provide an introduction to ways of analysing film and television, particularly Australian material. Analysing for thematic content and for encoded messages Mayer and Burton cover television advertising, soap operas and films such as *Crocodile Dundee, Man from Snowy River* and *Gallipoli.*

Webb *et al.* (1966) also look at cinema in their discussion of archives. They focused on the early work of Wolfenstein and Leites (1950) who analysed the 'good–bad girl' character in American films. A 'good' girl, because of circumstances beyond her control, seems to be a bad girl. The point is to reveal to the hero and the viewer, that she really was a 'good' girl all the time. The authors explore this set of images for an understanding of popular views about femininity.

In that tradition, Williamson (1986) continues with an examination of the film *Body Heat* and the films of Doris Day. Williamson compares her viewing of *Body Heat* with that of the publicity sheets. Looking at the issues of lust and illicit sexual desire and behaviour, she argues that desire becomes the improper and misleading 'culprit' in this type of film. In fact, the real problems are the primacy and reigning status given to social sanctions which are either constructed or reinforced by the film's images and organisation. Her analysis of Doris Day films is even more provocative. Williamson wonders why the golden virgin image of Doris Day is particularly fixed and difficult to shift, despite the fact that this is a very partial view of her actual work. She asks, is there a vested interest in forgetting those other roles which are counter evidence to the type cast? For example, in *Love Me or Leave Me* Day plays a nightclub singer raped by her lover (played by James Cagney). In *Storm Warning* she is 'married to a Ku Klux Klan member who shoots her and tries to rape her sister' (p. 147). Williamson makes a point worth remembering: 'Which image is "remembered" depends on what's in it for you' and the prevailing ideas of the time concerning family, sex, women, marriage and so on.

More recently, Sherman and Dominick (1986) have examined sex and violence in music videos and Norman Denzin (1991) has conducted analyses of films such as *Wall Street, Crimes and Misdemeanours, When Harry Met Sally*, and *Sex, Lies and Videotape*

Problems in analysing film and TV are:

- reliability (unsystematic methods, impressionistic)
- validity (multiple audiences make single pronouncements limited or invalid)
- subjective personal bias
- over-interpretation and intellectualising (the 'academic' message 'received' may not be the 'popular')

as a means of exploring dominant cultural ideas.

However, most of the above problems and concerns stem from a misapplication of methodological criteria. Many researchers in the humanities, particularly in the semiotic tradition, would not accept these as criticism. They might argue that reliability and validity are empiricist values which are part of predictive scientific models and which exist for the construction of unified theories. However, neither of those concerns is central to film and television criticism. Relevance is more important than reliability. How do media images reflect or become reflections of social processes? What symbolic and ideological baggage is brought, unpacked, projected, reinforced or suggested by media? The aim is identification and evaluation not simply arbitration.

Single interpretations are rarely offered as definitive or representative but rather as important (not necessarily prioritised) elements. These are characterising but not necessarily in any *total* way and may not be exhaustive or exclusive of other competing, indeed conflicting, elements. Of course, when something is argued to be true for some, this does not make it true for all. And, conversely, when something is not true for one audience this does not mean that it is not true for other audiences.

Furthermore, if an interpretation is valid for one person, it probably is so for some others too. Personal interpretations should not be viewed as constructions developed in a social vacuum by anomalous personalities. Like the personalities themselves, interpretations are perspectives forged from social experience of one sort or another. That fact gives personal interpretation both a political and a social significance, because all personal interpretations are shared by someone or other, somewhere, sometime. And the size and nature of that affinity is no measure of its possible influence or lack of it. Minority views and ideas may dominate or they may not. But these are other matters, decidable by other means. To label a view as simply personal or subjective is a bold attempt to dismiss that idea by declaring some views atomistic and asocial while other ideas are fertile and representative. The implication is that views which are less widely shared and less influential are therefore somehow less 'true'. However, representative or popular views are as transitory and complex as are non-representative ones and indeed unpopular views and styles of yesterday may be today's popular ones. The reverse may also occur. This is called fashion and change. Acknowledgment of this ensures that all interpretations are important because they are all politically and socially possible.

Finally, representative or popular views hide and silence unacceptable views. They discourage, dismiss or discredit alternative ideas. The development of personal interpretations then, helps 'invent allusions to the conceivable which cannot be presented'

(Lyotard, 1984, p. 81). This allows us to see the hidden and the possible alongside the dominating and the fashionable. That task becomes an important and vital methodological contribution because it actually rehabilitates the idea of representativeness. Representativeness can then be seen as a context and group dependent social process rather than in its usual association with superficial ideas about 'popularity' or 'dominance'. 'Representativeness' becomes a historical and changeable idea instead of one readily aligned to static notions of fashion and permanence. In that context the personal and the subjective are vital clues to a dynamic process rather than a deviant opinion which appears irrelevant to the social picture, analysis or portrayal.

The above comments notwithstanding, the variety of theoretical and methodological styles in film and television analyses does make a beginning analysis for the newcomer to the area fairly difficult. Fields (1988) provides a broad, practical framework for analyses which is both accessible and enabling. He applies this to an analysis of television news but the principles remain similar for media in general.

Fields' practical principles in analysing television

1 Unitise the content (break into sections such as speaker-to-speaker, shot-by-shot, sequence-by-sequence or scene-by-scene).
2 Transcribe (write down what was said and shown).
3 Code (tag and label analytic comment and own reaction, categories may emerge as you watch—in that case backtrack to apply these also).
4 Analyse the elements (expressions, scene, music, graphic, sound effect, verbal content—look for themes that recur).
5 Generate alternatives (look at other possible symbols, news versions, plot scenarios, to develop a sense of the omitted and the alternative).
6 Describe interplay of the elements (how the elements work together, for what effect. Look for themes—see Miles and Huberman, 1984.).
7 Explanation (review appropriate theories for cinema and television discourse).

The principles of thematic and semiotic analysis discussed in chapter three apply to the steps outlined by Fields (1988) above.

The importance of the visual and the sound dimension in audio-visual material cannot be overemphasised. Several writers warn that sight and sound have different functions in cinema and television. For example, Seiter (1987, p. 26) cites Ellis (1982) and Altman (1984) as arguing that sound dominates the television image where in cinema sound is much more subservient to image. Applause, program theme music and the speech of announcers tend to *precede* the image, calling a viewer's attention or presence back to the television screen.

To aid in this analysis of the different components or dimensions, Fields (1988) also suggests that watching a video tape without the sound and listening to the tape without watching are aids to watching and listening normally. He also suggests that photographing the television screen can help 'focus attention on the visuals' (p. 192).

Finally, Allen (1987) observes that traditional criticism of television news (or indeed commercial media generally) is that these are biased. But he goes on to argue that no presentation is 'unbiased'. The central questions are: How is the world represented? What are the particular themes or messages which are obvious and not so obvious? What influences lay behind these presentations and their agenda?. Despite the complexity of both media studies and human communications studies these are still some of the central questions for researchers and Fields' suggestions are a good practical start for your own attempts to answer them.

Music

The study of music has traditionally been the province of musicology, a field dominated by musicians and focused on the study of 'art music' or 'Western concert music' (though, of course, 'ethnomusicology' has included the study of 'world' folk music (Lomax, 1968; Nettl, 1956 and 1983; Hamm, Nettl and Byrnside, 1975)). According to Middleton (1990) and Shepherd (1987), musicology has largely avoided pop music because of an implicit or explicit elitism and a distrust for any closer relationship between music and business. However, when musicologists have deigned to analyse pop music, there have been other problems. Often their methods were biased and ethnocentric, privileging the taste and social values of nineteenth century Europe, especially German idealism (Middleton, 1990). The terminology used by traditional musicology tends to characterise pop music as both deviant and substandard. Like the early nineteenth century anthropological characterisation of non-Western societies, examples are described as simple, childish, raw, immoral or base. Traditional musicology is also notation (music score) centric and this downgrades performance and improvisation. Furthermore, the focus on texts particularly privileges the piano score and omits the drum, synthesised noise and special effects. Text/notation orientation is particularly unsuited to the complexities of modern, particular post-modern, music forms. Finally, Middleton argues that traditional musicologists ironically assume that music can be defined in a neutral, absolute and idealistic form away from social meaning, experience, argument and taste. This, of course, promotes a class-bound understanding of music which, apart from anything else, makes contemporary musicology rather narrow and anachronistic.

When dissident musicologists and sociologists of music have attempted to go beyond these problems through, for example, semiotic approaches, yet other problems have emerged. On the most basic level of unitising, of developing units for analysis, there have been anxieties over definition. Do you analyse the lyrics, the music or the listener categories? Is the song the basic unit or the genre? Or perhaps the social context is critical, that is, the party, dance or driving? Does a song's meaning exist outside its consumption context? Some argue that the only interpretation possible is an emic one. One must interpret music in terms of either the producer's intent or the culture of fans. But Pattison (1987) and Middleton (1987) are unhappy with the abstract 'idealism of traditional musicology and the reactionary materialism of those sociological reductionists who would argue that contexts rule'. Simple emic analyses ignore broader, larger scale continuums in music which are an important part of it. He argues, for example, that twelve-bar blues has distinct similarities with pazamezzo moderno and not nineteenth century hymns; and punk music is highly similar to sixteenth century dance music. These relationships should be part of any pursuit of musical–cultural explanation.

But despite these problems and reservations there have been many attempts to link music to broader cultural and social questions. Peterson and Berger (1975), for example, analysed the issue of industry concentration (monopoly) and market responsiveness and found that patterns of industry monopoly and change were more critical to music diversity than the pressures of simple market demand. They reviewed twenty-six years (from 1948–73) of pop music, charting the legal and economic changes to the industry alongside changes in music diversity marketed by the dominant companies. An important part of this analysis included content analysis of the music.

They argued that monopoly by the leading companies leads to homogeneity of product. This leads to stasis or even a shrinking market, as consumers get bored and look for alternatives, often live music sources. Changes to the laws regulating the industry led to a new influx of competitors. This in turn led to the signing up of 'new' artists and styles, from the pool of unrecorded artists locked out by the hegemony of taste regulated by the old monopolies. In the course of this analysis, Peterson and Berger show that between 1948–55, 80 per cent of songs fitted a formula love song format with little to no mention of social problems. From 1956–59, love themes were given more candid and personal treatment and frequently dealt with issues of conflict, especially with parents, school or work. The 1960s witnessed an explosion of diversity because of the restructuring of the music industry both locally and globally. From Beatlemania to California psychedelic love themes dominated but now in the context of 'broader social issues' (p. 167). Songs also dealt with sex, hypocrisy, politics, war and race relations. Various theories were developed to explain this: communist plots, sexual and drug induced decadence and 'white theft of black creativity' (p. 167) were among the most 'over-the-top'.

Ford (1971) studied styles of music to assess the role of geographic factors in the spread of certain types of music. He looked at how rock 'n' roll developed out of rhythm and blues and white country and western. This took him into a closer analysis of black gospel music and the chants of field hollers and cotton pickers. This field music was characterised not simply by beat and melody but also by the use of 'a statement by an individual and restatement by a chorus' (sometimes referred to as 'call and response'). He also traces the term 'rock 'n' roll' to lyric lines such as 'my baby rocks me with a steady roll' and to songs such as Bill Haley's 'Rock around the clock' and 'Shake, rattle and roll'. Moving away from music production to focus on musical presentation other studies such as Barnard (1988) look at how radio stations select and use pop music.

Recently Wollen (1986) argued that the music video is a good reflection of post-modern developments in popular music, television, advertising and fashion. Being highly eclectic and historical, its methods borrow, copy, parody, simulate, replicate and blend past with present, *passé* with possible. The recent (1992) diet

Coca-Cola advertisement which features Elton John 'shakin it up' with Humphrey Bogart, James Cagney and Louis Armstrong is a skilfully integrated video–music example. Through production and post-production techniques the present is blended with the past through allusion or direct 'quotation' to achieve a 'new', transcendent image of the contemporary and futuristic.

Continuing these remarks and extending them, Frith (1986) argues that these sophisticated examples of post-production technology have allowed old fears about 'authenticity' to escalate. The problem of 'real' vs. 'phoney' music and authorship has been with us for a long time. The use of the microphone, electric bands by folk artists and the use of synthesisers all provoked similar criticism and outcry, as do the current use of post-production special effects and backing tapes. However, Frith shows that that argument can be stood on its head. Far from indicating a betrayal of the idea of a personal entertainer's 'direct' and 'honest' relationship with listeners, current post-production technologies can increase the creativity and individual expression of artists. Furthermore, art ownership may become democratised as recognition spreads that not one, but several 'artists' contribute to the final effect. In the case of the diet Coca-Cola advertisement, not only Elton John but also Bogart and a host of engineers, are the 'artists' and 'owners' of the product. This, as the post-structuralists are fond of saying, 'decentres' the idea of totalising concepts of ownership such as copyright. Copyright is the linchpin of power and control in the music industry and post-modern developments can therefore be seen as critical of these old concepts of ownership. These ideas of ownership are also associated with the idea that the more sophisticated the act, the less the actor is involved because of the element of trickery. Furthermore, the idea that musical pieces are final products rather than pieces that can be remixed or added to and recombined with others (work-in-progress) bring further claims about the creative bankruptcy of post-modern musical styles. These sentiments and criticisms, according to Frith, however, show how strongly both producers and some audiences are attached to a nineteenth century notion of originality, authorship and entertainment.

The examples discussed above illustrate how an analysis of genre, of popular musical style, can lead to comments about broader cultural processes. However, there have been many studies that have attempted to be more focused and specific, and have concentrated on the task of interpreting songs. This practice has quite a few problems also.

First, a concentration on the connoted meanings of music (ideational, musical associative and evocative) ignores the fact that these are *secondary* features of the music. As Williamson (1986) argues, people bring things to images, they project onto images, as well as receive. In other words, music does not simply *send* messages to listeners, it also *evokes ones already existing* in the listener. People *bring* feelings and ideas to music as well as receive them. Music often signifies beyond itself but not necessarily to 'the outside world'. It may signify to the genre and style to which it is a part (Middleton, 1987). Content analysis is therefore hazardous because it can decontextualise this connection. Complicating this is the complexity and variability of different styles of music.

Middleton (1990) reminds us that in some styles of music or song, the words are designed to augment the music, they are performing quasi-musical functions complementing the instrumentals. In other music, such as the political message song, the words are of primary significance; the music mere support. In ballads 'verbal dominance is less clear cut' (p. 229).

Finally, over-reliance on a simple content or semiotic analysis of songs can be very misleading. Robinson and Hirsch's (1972) study of American high school students and their listening habits is a wonderful illustration of this point. They found that listening habits were class bound. Protest songs, for example, were favoured by middle-class children while general pop songs were favoured by working-class children. Black children favoured rhythm and blues in a ratio of 8 to 1 over white children. Liking a song and knowing its lyrics, however, are two separate issues. Most people (70 per cent) could not remember lyrics or interpret them—most were simply attracted to the sound or tune. Listeners tuned in or out of their radio station depending on whether their style of music was being played. (Not everyone agrees with these findings. See, for example, Pattison's (1987, p. viii) remarks on the subject.) Listening groups are not homogeneous.

As with all analytic work of audio-visual materials it is best *not* to solely rely on a content or semiotic analysis. It is particularly important to use other sources of historical and cultural knowledge about the music. It is also worth remembering that the research designs discussed in chapter two all emphasise the importance of prior reading and this may enable budding musicologists to identify listener responses, artistic design and the social features of the audience before and during the process of any text analysis. Beware of simple lyric analysis, simply because the lyric may not be signif-

Problems in analysing music

- Reading of music style and lyric may not be generalisable to specific audiences without first knowing who the audience is (men/women, black/white, blues/white collar etc.). Emic/etic problem no. 1—social features.
- Reading of music style and lyric may not be generalisable to specific audiences without knowing the purpose to which listeners are using the music (dancing, driving, making love, contemplation). Emic/etic problem no. 2—social circumstances.
- Reading of music style and lyric may not be generalisable to specific audiences without knowing the design features of the music (producer's artistic intent, functional style), for example rap, ballad, message, mood music. Emic/etic problem no. 3—artistic design.

icant without the sounds and, in any case, it may not be the most influential stimulus to a listener. Become aware of important associations in your choice of music. Stefani (1973) quoting Middleton (1987) mentions several important ones for pop music.

Stefani's pop music associations

- Intentional values
  Synthesiser = technology/modernity
  Boogie = erotic
- Ideological values
  Psychedelic songs = drugs
  Country music = conservative politics
- Emotive values
  Punk music = aggression
  Singer/songwriters = soulful, confessional
- Style values
  Rock 'n' roll = bikes, violence, jeans
- Moral values
  Rock 'n' roll = corrupt/liberating
  Classic = uplifting/spiritual/intellectual

The important issue is deciding which, if any, of these terms of reference applies to your interest and, furthermore, to identify to whom these values apply—insiders' meanings (fans, followers and audiences) or outsiders' (critics, detractors and rivals). Finally, avoid stereotyping through uncritical applications of Stefani 'values'. These categories should alert you to possibilities, not necessarily define them.

Finally, don't be afraid to socially analyse music, even given the problems identified above. Identifying problems should sharpen rather than immobilise your attention.

Practical principles for analysing music

- Listen to music carefully and repeatedly. Hood (1971) provides some practical suggestions here:
  - a) Listen to music you love and music you dislike, returning to each regularly to explore the reasons for your preferences, to sensitise your ear and to liberalise your ideas about music.
  - b) Attend live performances where possible because all recordings are selective in their technological processing and receiving equipment.
  - c) Learn to sing or play some instruments yourself. This can quickly give you a different and more sensitive perspective on performance and musical nuance.
- Collect opinion and interpretation as you would book references, noting and considering them all in both playful and serious ways.
- Familiarise yourself with the history and sociology of your chosen piece/pieces of music.

Henri Lefebvre (1971) and Richard Middleton (1990) argue that music is where fantasy can be realised. It is where individuals can rearrange the everyday habits of thought and feeling and construct, however fleetingly, a subterranean world of private fancy. Music mediates culture and is itself constituted by it. But more than this, music affects the body more directly than do simple optic images. Music reflects and courts the body's own heartbeat, its breathing, its electricity and wavelengths. Music has a physical relationship with us. Its power therefore, is unconscious as well as interpersonal, erotic as well as abstract. Avoid therefore, an intellectual tendency to sanitise a musicological analysis by confining one's interpretation to the technical and social. To exclude feelings and their attachments from any musical study is to depersonalise and prevent the unconscious from contributing its serendipity. This is a process which Middleton reminds us also removes the body from its central role in musical production and enjoyment.

Ethics

Ethical problems concerning the use of audio-visual material can be very subtle. The ethical issues surrounding photographing people are discussed in the chapter on hardware and software. If, however, you are simply using photographs or film that other people have taken then, clearly, permission should always be sought. Remember that permission to view the photos and to use them in any analysis is *not* the same as gaining permission to display them publicly (for example, in a book or exhibition). Permission will need to be sought for this and provisions made for their *care, copying* and/or *return*. If you have someone's photographs or films remember that ethical behaviour extends to these issues too, and not simply to consent alone. In this respect, take care to ascertain people's attitude to photographs. Many Aborigines, for example, do not wish to see images of dead relatives or even have the deceased person's name mentioned (Elkin, 1945, p. 238).

Remember also that when commenting and analysing photographs and film footage that you are not commenting on 'mere history' or 'social processes'. Some living relatives, or indeed the subjects of the images themselves, may still be alive and can be offended or upset by unreasonable or insensitive commentary. The fine line between the necessary and the gratuitous is both thin and ambiguous and is more readily apparent to the litigant than the offender. Take care and beware not to defame. With film and music, particularly if you wish to reproduce or display, do take care to fulfil all the copyright obligations to the owners. You might consult your local library for current copyright regulations governing the reproduction of published photograph and film material. Alternatively, you can phone the Australian Copyright Council on its free '008' number and ask for information about current rules and regulations. This is a free service to the public which offers legal opinion and sends published material to the inquirer.

Finally, Bruno Nettl (1983, pp. 290–300) reminds us of some of the ethical issues in collecting and studying other people's music, particularly the music of the poor and marginalised. He discusses the problem of researcher benefits (publishing, royalties, prestige, promotion, etc.) which do not flow on to those one has studied. Also the portrayal of a group's music may lead to fear or hostility from others towards that group. Nettl argues that perhaps one's informants or the community from which the music is studied can be recognised and empowered as co-investigators, even co-authors. This raises, once again, serious questions about the purpose of research and the politics of research ownership, issues which are not necessarily able to be bypassed unobtrusively.

Recommended reading

Allen, R.C. (ed.) (1987), *Channels of Discourse*, Routledge, London

Burton, G. (1990), *More than Meets the Eye*, Edward Arnold, London

Curry, T.J. and Clarke, A.C. (1983), *Introducing Visual Sociology*, Kendall/Hunt, Dubuque, Iowa

Fields, E.E. (1988), 'Qualitative content analysis of television news: systematic techniques', *Qualitative Sociology*, vol. 11, no. 3, pp. 183–93

Hood, M. (1971), *The Ethnomusicologist,* McGraw-Hill, New York

Mayer, G. and Burton, L. (1991), *Media Studies,* Jacaranda Press, Milton, Qld

Middleton, R. (1990), *Studying Popular Music*, Open University Press, London

Nettl, B. (1983), *The Study of Ethnomusicology: Twenty-nine Issues and Concepts*, University of Illinois Press, Urbana

Walker, A.L. and Moulton, R.K. (1989), 'Photo albums: images of time and reflections of self', *Qualitative Sociology*, vol. 12, no. 2, pp. 155–83

Williamson, J. (1987), *Consuming Passions*, Marion Boyers, London

6

Material culture

In 1966 when Webb and his colleagues wrote about the study of the physical they confined themselves to 'traces'. To illustrate what they meant by that term they began their chapter with a Sherlock Holmes story. Holmes congratulates Watson on his purchase of one of a pair of professional office suites. When asked the basis of this, Holmes points to the fact that the steps leading to Watson's suite were more worn, implying more profitable people traffic to this suite as compared to its twin on the other side.

Physical traces were evidences of wear or erosion, but also of accretion, of things created or added. For traces of erosion, Webb and his colleagues looked at studies of floor and step wear, of wear to door handles and even statues. They described the value of studying dog-eared or soiled pages of books as indicators of popularity and use. In accretion studies they described research which checked the radio dials of cars brought in for service, of suits of armour (to assess the height of medieval knights), the contents of garbage, graffiti and street litter. Some of these studies remain popular, particularly garbage and graffiti analysis. But Webb and his colleagues did not broaden their brief to include the study of objects and organisational spaces. Insights from architecture, archaeology or food preparation and content for example, were ignored or given scant attention. Without including these areas they indulged in some far-out speculation. For example, the authors suggested a study of 'noseprints on glass fronted exhibits' arguing that height of print will be associated with age! (as if everyone would dob their nose on a glass exhibit or not bend to examine some object lower and more closely!).

In this chapter, I will overview some very interesting studies of the old type, graffiti, garbage and cemeteries, because these remain important sources of cultural information. But I will also mention some less obvious studies of objects and relations that we take for granted in everyday life—the motor car or the chair. Objects and traces have as many ways of studying them as the objects themselves dictate. The specific principles in studying house styles, for example, are perhaps different from those needed in a study of smoker's pipes, and certainly different types of homework and observational skill may be required. Unlike archival work or film analysis, there is no limit to the number of physical settings, objects or traces one may study. So, unlike other chapters, I will spend longer reviewing and summarising past studies to demonstrate something of the range and diversity involved.

So, given that complexity, what follows is a quick overview of some fascinating studies that might serve as a rich source for your own ideas.There will be an outline of the main advantages and disadvantages and, finally, some principles for general analysis. These principles, when combined with your favourite technique for pattern recognition, should enable you to analyse most physical settings, objects and traces. The chapter will end with a discussion of ethical issues and questions.

Some studies

Rathje (1979) begins his discussion of physical traces in a similar fashion to his mentors and predecessors in *Unobtrusive Measures*. He begins with a story from Sherlock Holmes. Here, Conan Doyle argues through his astute sleuth that the scratched keyhole in men's watches is a sign of one too many alcoholic beverages late at night. That example of what wear can tell us introduces Carpenter's (1977) study of screwdriver use. Do people only use screwdrivers to turn screws? The study compares what people say about usage with the physical examination of the tool itself. People say that they drive screws but the chipping, bending and paint marks on most screwdrivers suggest other uses—as chisels, stirrers, can-openers, knives, digging implements and so on.

Rathje also discusses the content analysis of garbage from houses and the local tip, a pastime for which he is very well known.

He discusses sexism on headstones and attempts to link income level with lawn clipping output.

Deetz (1977) discusses the impact of the Renaissance on early American social life as revealed in objects as diverse as ceramics, house designs, gravestones (a re-occuring obsession of people in this area), furniture and cutlery. For example, he discusses the recent historical emphasis on individualism by demonstrating how this is reflected in the individual portioning of meals. He discusses the gradual democratisation which emerged from the disintegration of the medieval political system by studying the proliferation and design of chairs. The chair, previously reserved for those in authority and leadership before the sixteenth century (the plebs sat on the floor), became first the preserve of the middle class, and then everyone's.

In Gould and Schiffer's (1981) fascinating book on the *Archaeology of Us*, several contributors demonstrate how much can be understood from a closer look around us. Rotheschild (1981) for example, studied pennies, in this case American one cent coins, and analysed the 'facial' characteristics in the light of theories about money, ritual, exchange and production. Cleghorn (1981) looked at the contents of a community store in the country, charting the hidden international links such a place has (because of the origins of the products), revising our notion of 'isolated' rural communities. Finally, Bath in the same collection, went a step further and looked at the organisational spaces of a modern supermarket, its shelf content and floor plan. Bath examined the various oppositional symbolic elements to discern their wider meaning—raw/cooked, wet/dry, oven ready/table ready and so on.

More recently, and less archaeologically, Clegg (1990) in a wide ranging study of organisations in the modern world critiques current sociological perspectives on this topic. His critique takes him to examine some unusual topics for a political and social theorist—the production of French bread and the study of Italian fashion.

More in the cultural studies tradition than archaeology or sociology, Bayley (1986) discusses the symbolism of the modern motor car. He examines its broader meanings, drawing primarily from literature, history and the cinema. He argues that the car has become an important symbol of material success (status symbol) but is also a machine which produces 'sensations of power or control'.

For some people, the car is a liberating symbol of personal choice, self-expression and destiny. In cinema, cars have been portrayed as tragic, menacing, cutesy, hostile, erotic and supernat-

For some people, the car is a liberating symbol of self-expression. Courtesy of the *Geelong Advertiser*

ural. In early cinema, motor cars were symbols of hope, progress and prosperity. The car is one of the first major modern products that had an explicit marketing policy of selling *appearance*. According to Bayley, Du Pont chemicals developed coloured synthetic paint call 'DU-CO' to replace the black of early Ford cars and the race for diversity and style began in earnest. The use of modelling clay instead of wood drove the design towards curves. The obsession of early designers with the military and aeroplanes gave cars in the 1950s and 1960s fins, flaps and scoops. Front grills became heraldic and sexual symbols and thus became the focus of industrial competitiveness and the measure for evaluating success or failure with consumers.

Bayley (1986, p. 20) documents a lawyer writing to complain about one of Ford's latest models. The complaint highlights the obviousness of some of the design imagery for some consumers:

> It was bad enough that Studebaker saw fit to design a car whose front reminds me of male testicles, but now you have gone that company one better by designing a car with a front like a female vagina.

In a similar vein to Bayley, Willis (1978) discussed the motor bike and its symbolic importance to bike gangs. His was a participant observer study to be sure, but the study of the central sentimentalised object of the gang was a critical part of this. Bikes, argued Willis, 'underscored' the personal and social identity of gang members, particularly in areas of roughness, masculinity and aggression. The style of bike and the garment of its rider convey wildness and intimidation to the public through integrated use of leather, studs, cattle-horn handlebars and extended front-wheel forks. Double chrome exhausts, mud guards and the removal of the exhaust silencer enhance that image. The mechanical qualities are recognised and shaped into human qualities.

These are, in their turn, modified further into yet other mechanical and social meanings. The bike as production object is modified, personalised and 'made over' into the cultural image of its owner and his culture—it is as Miller (1988) puts it, 'appropriated'.

Miller interviewed forty tenants in a public housing estate and then photographed most of their kitchens. He was interested in how people overcame a given, foreign environment and made it their own through personalising practices. Reasoning that owner building will not be the province of most people, he sees housing as a consumption rather than production issue. In public housing in particular, consumption is a process of transforming alien, 'not me' environments into familiar, personalised ones which reflect social and biographical tastes and values. Who transforms kitchens in this way and what might it tell us culturally?

He found that a small group of kitchens remained unaltered and, at the other extreme, there were kitchens which had been ripped out and replaced by commercially built ones. Although kitchen company advertising showed young people with new laminated kitchens and older people with 'olde-worlde' wood-style kitchens, the experience of consumers appeared to be the opposite. For the young, the laminates looked old because laminates have been around a long time. They therefore preferred the woody ones. For the old, the laminates looked modern and these were preferred. West Indians preferred heavily cluttered kitchens painted brown, orange or purple while those of Anglo-Saxon origin preferred pastels and covered their kitchens with memorabilia. The latter also dotted their kitchens with bread boards, tea towels and cosies and trays, particularly trays and tea towels with bold flowers, dogs or cats on them.

Miller links the kitchens to broader patterns of sociability, to socialising. The more people socialise the more they attend to their kitchens. In that context lonely, isolated or depressed people tended to accept the public housing kitchen unchanged. Others, particularly single women, decorated their kitchens with a barrage of personal objects. Families with men were more likely to physically alter or replace their kitchens.

Home interiors were also photographed and studied by Togashi (cited in Collier and Collier, 1986, p. 199) who wanted to look at cultural persistence in American–Japanese immigrants. She photographed the interiors of first, second and third generation American–Japanese households to assess the persistence of Japanese culture in these groups. But not everyone who is interested in houses as objects studies rooms. Vera (1989) wrote a fascinating article documenting his study of Dutch windows. The Dutch leave their windows uncovered during the day and night. The windows are heavily decorated by lace curtains, personal objects, posters and/or plants. They permit people to observe each other in a game of not looking like they are looking at each other. Vera describes how this uncovered window approach allows public areas in residential areas to became quasi-private and how certain parts of the inside of a house become quasi-public.

But the really interesting aspect of this study is its methodology. First, it is an observational study of a physical object but it also employs photography, discussion sessions with colleagues in social sciences and architecture, informal chats with residents, window sellers and even museums. The approach then, is somewhat ethnographic, but the physical object is the primary focus.

Second, Vera sweeps aside the usual hypothetico–deductive and ethnographic–inductive approaches to employ a more experimental style. Rather than 'test' a theory or 'develop' one or two theories from a thematic analysis, Vera explores a variety of ideas suggested to him by others or those which simply occur to him through association. He does not privilege any one idea but entertains all of them as cultural possibility. He entertains historical, sociological, commercial, architectural and personal ideas about the origins and functions of the window *not* as final explanations but as 'conceptual frames', each in its own right a potential direction for sociological (or any other kind of future) research.

His techniques of pattern recognition were a combination of content and thematic analysis. His theoretical discussion draws from oral suggestions from acquaintances (lay and professional) as well as the literature, such as it is, on windows. As I mentioned in the chapter on audio-visual sources, this is also an excellent way to approach the study of music. Collect opinion and interpretation in exactly the same way as you would references or texts from books. Consider all you collect in the light of the literature. For social practices (such as music) which may have many meanings or whose meanings may have endless complexities, cast your net of theoretical/interpretative possibilities as wide as possible. Vera's study of Dutch windows is a fine example of just this technique.

From gross objects to smaller traces: there have been many interesting studies of graffiti. Blake (1981), for instance, looked at graffiti in male toilets at the University of Hawaii. He was interested in inter-racial content. And Singleton and his colleagues (1988) describe the earlier study of Bruner and Kelso of restroom graffiti. This was a semiotic study of same sex communication. Bruner and Kelso argued that women's graffiti indicated that they were interpersonal and interactive compared with men. Men were more egocentric and competitive. Women offered each other advice or raised questions while men were more insulting and condescending. The differences in the graffiti styles can be attributed to the respective political positions of the sexes. Men write graffiti to tell themselves and others that they are in control. Women write co-operatively in the style of those united by their mutual subordination.

Klofas and Cutshall (1985) provide a useful review of studies that use graffiti and for anyone interested in this type of study, the review leads them to work done in the ruins of Athens and Pompeii. Klofas and Cutshall, however, write primarily about their study of graffiti in a juvenile detention centre. While most graffiti in public comprises messages or obscenities around the theme of sex, nearly 50 per cent of graffiti in this study comprised personal identifiers (names or initials) with their home town or State. Only twenty-nine specimens from a total of 3000 graffiti traces mention sex. And this highlights the significance of group affiliations *prior* to entry to the centre.

Rathje, as I mentioned earlier, is also well known for his content analysis of household garbage, but he's not the only one interested. Phil Murphy, the son of Dan Murphy (a famous Australian wine and beer seller) was reportedly seen rummaging through the rubbish bins of a Melbourne suburb. This was the same area in which he was about to open a shop—a bottle shop. 'Clever Phil was getting a feel for what the locals were drinking' ('Spy Hole', *The Sunday Age*, 8 March, 1992).

Studies of garbage are not only useful to business—the Environmental Protection Authority (EPA) is also interested. They have discovered that if you give a person a bin the size of your coat pocket, they'll fill it to the brim. No surprise of course. However, if you give them another bin the size of a small block of flats they'll do the same. People fill bins no matter what the size.

The EPA found that garbage varied from municipality to municipality but was rather consistent within one. Working-class areas prefer beer in cans and stubbies while middle-class areas disposed of many more wine bottles. Other suburbs showed their love of dogs by the number of dog food cans (or their preference for dog food in cans!). Some areas revealed social values of a more interesting nature.

Dr Brian Robinson (Chairperson of the EPA) was quoted by *The Age* (24 November, 1991) commenting on bottle recycling bags and social class. Working-class people were more shy in filling their bags for fear that 'people might think that they had an alcohol problem, whereas people who lived in yuppie areas filled the bags up to show how environmentally sound they were' (p. 5).

Not everything that we are finished with, however, ends up in the bin at the local tip. There has been a long interest by social researchers in cemeteries. Webb and his colleagues classified tombstones as archival but many other researchers, particularly those whose interests are broader than simple epitaphs, see them also as physical art work. Irwin's (1981) is merely one of hundreds of examples of studies which look at crypts and tombs. Irwin was interested in eighteenth and nineteenth century tombs in France, Italy and England. Looking at design, style and iconography, he developed a discussion of attitudes to death in those periods and linked these to ideas popular at the time in theology, literature and art. But not everyone who studies cemeteries is interested in death. Dethlefsen (1981) sees it another way—as a way of studying the living and their attitudes to life.

Dethlefsen documents his travels through American graveyards and makes two very good cultural points: first, the cemetery is a place designed and organised by the living, from their cultural values and practices; second, cemeteries are good starting points for asking general questions about the community. We will return to this last point towards the end of the chapter.

The cemetery, as landscape and as 'text', can be used to inductively develop theories about the cultural life of a community. Indeed, this is what Dethlefsen does in his own article, escorting the reader through different periods of American history. A year before Dethlefsen's study appeared, an Australian historian and genealogist published a book conducting a similar tour of Australia.

Gilbert (1980) in his ironically, but humorously, entitled book *A Grave Look at History* suggests systematic ways of examining Australian cemeteries. He categorises grave styles by:

- gravestone (upright slab, obelisk, statuary, vault etc.)
- letters and script styles of the writing
- epitaphs (which he divided into biographic details, messages of the dead to the living, messages of the living to the dead, and emotional expressions of loss, resentment, hope, affection etc.)
- rails and pickets (to stop stock from scratching themselves up against the gravestones)

Additional to these categories Dethlefsen (1981) adds:

- pictures, decorations and ornaments around or on the grave
- the geography of the cemetery itself, that is, the social divisions and locations of various religious denominations or special types of graves (for example, military).

Any study of a cemetery should combine with other sources of information: 'newspaper obituary notices, mourning cards (issued by undertakers for the mourners), death certificates, undertakers records and church registers' (Gilbert, 1980, p. 126).

Gilbert also warns of dangers. Like archive research, loping around a cemetery has some practical hazards. Look out for snakes (in Australia). Be careful around tall, heavy monuments for many of them have slight foundations or have had their mortar badly weathered over time. Do not lean on anything and most certainly do not rest in the shade of one. Many have tumbled down without warning. Research at a local cemetery should be a temporary affair! Also watch for nettle, blackberries, lantana and briar roses. These can make a mess of your legs or your clothes. Gilbert advises that you take a companion for an extra pair of eyes and for assistance in general. Apart from this, Gilbert also supplies a worksheet for budding researchers to document the physical, literary and demographic patterns in cemeteries. This is very useful. If you are planning to do a bit of cemetery research, don't leave home without it.

Overall, this quick overview has suggested a variety of objects, settings and traces one might research. Let me provide a summary list to consider—before moving on to discuss the strengths and weaknesses of work in this area.

Research possibilities	
steps	motor cars
floors	motor bikes
door handles	chairs
statues	pipes (smoking)
books	watches
book pages	screwdrivers
radio dials	ceramics
suits of armour	house designs
garbage	gravestones
graffiti	cutlery
street litter	money
museum exhibits	shops/supermarkets
French bread	Italian fashion
kitchens	interior decorations
windows	recycling bag contents
oriental rugs	lawn clippings
clothing	Chinese food

Problems

There are four main advantages associated with the use of physical settings, objects and traces. First, material evidence can supply data or information about groups which either do not, will not or are unable to supply some of this in oral or written form. Second, they are valuable when assessing the extent to which an activity has occurred. This applies particularly to physical traces of wear and some accretions such as garbage. Third, they provide certain types of historical and social data that traditional methods could not. The study of screwdrivers is instructive here, as indeed is the study of Dutch windows. Fourth, trace analysis in particular can provide a minimal estimate of activity. In other words, although traces may underestimate activity they may nevertheless provide a conservative estimate. We know that at least this or that much activity has taken place. Other activity may have occured which is not detected but at the very least the trace records the minimum level. This is Rathje's argument for the study of garbage. Finally, physical settings, objects and traces can be valuable sources of questions about culture, its social institutions and practices.

These are not simply 'historical' questions. Some may also be surprisingly contemporary in nature. For example, a cemetery quite close to my university had an unmarked area filled with five hundred stillborn foetuses and children who died at birth. The cemetery had an 'arrangement' with local hospitals to bury them in this unmarked spot. Some ten years after this practice ceased, the mothers returned to the hospital and then the cemetery in search of the whereabouts of their children. Now, half a dozen plaques 'memorialise' the place where these children have been buried. What has happened to community attitudes to the stillborn in that time? How have women's attitudes and society's attitudes changed in the last two decades in respect to motherhood, childbirth, children and death? Is this practice of 'memorialising' increasing? How, if at all, are our social institutions (such as medicine, funeral industry etc.) responding to these changes?

The study of cemeteries is an obvious example but the study of cars or kitchens can also challenge one's usual expectations and be richly suggestive.

The major disadvantage is the selective survival of materials. Depending on a host of social, historical or geographical conditions objects or traces may be erased or vary because of unknown intervening variables. This of course, raises the whole question of

Advantages of studying material culture

- An alternative source of data about groups which do not provide written or oral evidence about their lives
- Provides an assessment of the extent of an activity
- Provides unusual types of social/historical data
- Provides minimal estimate of activity
- Suggests questions about broader cultural processes

how representative are objects or traces? Second, traces may indicate the extent and type of activity but they do not always indicate the characteristic of the population (Webb *et al.*, 1966). Lack of certainty surrounding survival of material may compound this problem. Third, there are the usual allegations about emic/etic problems. You say the object or the setting has this meaning but is this the actual meaning embraced by the users. How do you know without asking? Finally, Rathje (1979) argues that, especially for traces, erosion and accression measures may distort, overlay or eradicate other ones. This makes constructing an idea of successive processes difficult if not sometimes impossible.

Disadvantages of studying material culture

- Selective survival of materials (representativeness?)
- Does not always indicate characteristics of the population
- Emic/etic problems
- Distortion or eradication of other forms/processes

Most of these disadvantages can be overcome or minimised by remembering to supplement your methodology with other work—archival, observational, conversational, literary. The study of physical things alone provides only rough and minimal estimates, however valuable.

However, the emic/etic issue is not necessarily surmounted by advice to supplement your observations and it is to that problem that archaeologist Hodder (1989) turns his attention. In a typically post-modernist paper entitled 'This is not an article on material culture as text' he proceeds to deconstruct the emic/etic concern in archaeology.

Hodder argues that archaeologists have privileged the idea of emic meaning as ideas which are part of conscious, intentional social communication contexts. If you want to know why people do things, you ask them, assuming on your part that they themselves actually know or are consciously aware. Archaeological work has always been seen as etic work on the simple assumption that you could never interrogate the original inhabitants.

Hodder critiques these sentiments/assumptions from a post-structuralist position. He is particularly influenced by the work of Ricoeur (1971). From this perspective, he argues that material culture should not be seen as the by-product of readily agreed upon ideas, but rather changing ideas based on discourse. This discourse is the subtle exchanges of ideas about one event, object, practice or whatever. Ideas about one thing constantly change and we may read/interpret these processes as text, as a book or story. The story changes a little every time we read it because the time or circumstances in which we read it are different. As we change, 'the story' changes in subtle ways. This is an exchange between reader and text that is not always conscious.

Ultimately this means that material culture, just as with written texts, can always be re-read or reinterpreted away from the intention of the authors. In this sense, interpretations have an arbitrary nature about them. To control this tendency to arbitrarily interpret meanings, writers (and all other people in society) attempt to constrain the potential range of meanings. Hence there may be many interpretations about what I mean by writing this book but it cannot be construed that I am writing a manual on kitchen plumbing. However, despite my attempts (through use of words, style and content coverage) to control for one set of meanings, others may be developed beyond the range of my conscious intention.

My intention may be to design a reasonably user-friendly introduction to unobtrusive methods of social research. Others may see it as an attack on the methodological hegemony and orthodoxy of the interview/survey in social sciences and humanities. That latter suggestion is not my intention, certainly not my major interest in writing this book. However, it is not a bad idea I think, and certainly not a foreign idea. In fact, if it became popular I might speak of the book in that context ('colonise' the idea'—make the idea my own, or 'appropriate' it) and in that case the etic becomes the emic (your idea becomes mine). It is no longer simply a question of *my* intention and *your* interpretation. Ideas about the nature of this book are discursive, interactional, dialectical—we feed off each other, renew each others ideas about each other. We adapt. As Hodder (1989, p. 255) argues, seeing material culture as text allows us to see that 'etic material culture might contribute to emic meanings' and that 'the meaning of language might be situationally adaptive'.

Physical objects, particularly if contextualised, provide opportunities for people to reread meanings. This process is not unlimited but, as in this book example, is constrained by context. The more complicated the behaviour (for example, royal coronation ceremony) or object (for example, sewing machine) the easier it is to discern its general meaning. Complexity conveys a specificity of function. The simpler the behaviour (a smile) or object (screwdriver) the more numerous the possible uses and function and hence interpretation. A motor bike will never be mistaken for a box of tomatoes, even by future archaeologists. But, if they interpret the bike as a twentieth century object which was designed to critique conformity, who are we to say that they may not be, after all, correct in that view. This is because people do not always give, or understand consciously, the reasons for their preferences any more than those who have highly erudite theories about them. Material culture is not simply a by-product of ideas but it is also the stimulant to ideas and our interpretations are part of that continual cycle of meaning.

Spooner's (1986) study of oriental carpets is a further good example of this complexity. Many people who are attracted by oriental rugs worry madly about the issue of authenticity. However, the original meanings of the art and design have largely been forgotten by the tribal producers in Iran, Pakistan and elsewhere. The idea that an object (i.e. the rug) must be 'authentic' (i.e. a genuine ethnic symbol of tribal and religious life) is part of the Western buyer's pursuit of the unique, the non mass-produced product. The 'native' handicraft reveals both authenticity in the honesty of human labour and individuality in the buyer's self-image as a discerning connoisseur. Carpets are used to negotiate not just the relative status, but also the quality of personality, or how one should be understood and appreciated as an individual by others. That search for authenticity helps construct our individuality. For rugs, however, the more we search for 'authenticity', the more the manufacturers frustrate that search by adapting their production and designs to satisfy it.

This is very similar to the story of Chinese food in Australia. Chinese food in the 1960s in Australia had the same status as the local fish and chips shop, it was 'down-market', low status food. Eating at a Chinese restaurant was not really 'eating out'. To 'eat out' one dined at French or even Italian restaurants. As Australia became more educated and better travelled in the 1970s and as Asia became more important to us, politically, socially and economically, we changed our view of the humble Chinese restaurant. A culture of the Asia-wise Aussie sprang up. We began to peep inside those restaurants and look for Chinese faces, the reasoning being that, if Chinese ate there it must be 'genuine', 'authentic'. This was the beginning or a feature part of the 'great search' for 'real' Chinese food. Woks began to sell like dim sims the decade before. Chinatown in both Sydney and Melbourne experienced a flood of eager Anglo-Australian connoisseurs. We were educated by stories of what 'real' Chinese food was from friends who had recently eaten in China during their cycling holiday there.

Sweet and sour pork and beef in black bean sauce were declared the 'foods the Chinese fed the whites' (read: those who aren't in the know). Sweet and sour pork came to know character assassination. In turn, the Chinese food business elite, watching all this rather closely, developed a new set of restaurants—boasting provincial foods or 'up-market' hybrids—for this educated and travelled clientele. For the connoisseur, Chinese restaurants had to be rather grim looking, with laminex tables, surly waiters, and room dimensions so small that you could hardly wield your chopsticks. These should also have naked, red ducks, chickens or pigs hanging upside down in the window and have 'ethnic' names like Gin Doys, Fong Loongs or Tan Blam Nam (no longer House of the Golden Sun!). Other restaurants catered for those who blanche at the idea of a 'head on' confrontation with authenticity. They have more room, lots of wood and bamboo decorations, friendly service in black and white attire, fish tanks and either 'ethnic' music or Vivaldi playing in the background. They have names like Madame Butterflys, Ming Dynasty, The Lucky Duck or they may even have a trendy New Yorker name such as Jimmys or Honkers or, as one Tasmanian restaurant, The Orient Express (a take-away joint).

So, just as Anglo-Australians change their idea of the 'authentic', Chinese business is right there, changing along with them. The suppliers (Chinese restaurants) and the consumers (those that 'don't know' and those that 'know') are constantly locked in discourse over mixed and interchanging meanings about what is 'genuine'. This is because most consumers rely on what is presented to them for their assessment. It is also because the suppliers themselves are often not regional peasants who bring their home cooking to sell, but simply Malaysian, Thai or Hong Kong urbanites trying to work out what their Australian and fellow urban Chinese want to eat.

The discourse about 'authenticity' is not about authenticity at all. Negotiating the meaning of food in this context is about identity construction on the one hand, and market appraisal on the other. It is not the meaning of one or the other which is important but rather the ongoing discourse *between* them. And this, like the meanings of all physical objects during and after their use, is an ongoing process. The emic and the etic are not opposites but rather reflections of each other.

Questioning the physical

The study of physical objects can supply basic, if minimal information about peoples lives and that is valuable in its own right. However, as Dethlefsen (1981) remarks about cemeteries, objects can be starting points for the asking of general questions about the wider culture. Often one can 'read' objects for the areas of life to which they refer, suggesting that you might interrogate them.

- What physical objects or traces do people create to express themselves? (Remember the graffiti studies.)
- To what extent and how do users relate and express themselves through what parts of the design and functions of an object? What meanings are accorded to shape, colour, feel, sound or spacial relations of the object/s? (Remember the EPA's finding on recycling bags and Dethlefsen's study of graveyards.)
- How, if at all, are these designs or relationships comparable to development in design in other objects around the same period? How, for example, does chair design relate to food customs of the time, or car design relate to aircraft design or house design? (Remember Bayley's study of cars and Willis's study of bikes.)
- How is the 'not me' made 'mine'? How are impersonal environments and objects personalised? (Remember Miller's study of kitchens.)
- How might objects mediate discourse? What variety of interpretations are stimulated by the object and its relations (real relations or implied)? (Remember the oriental rugs and Chinese food example.)
- Do you think the object or setting has or had only one meaning for insiders (emic)? If not, how many?

Practice principles for analysing the material

- Research the background/secondary literature.
- Solicit a wide range of professional/lay opinion.
- Use complementary/supplementary methods—archival, observational, conversational.
- Discuss questions that the object raises:
  - —What does this say about users/creators?
  - —What role does design play in this?
  - —How are these shapes/forms part of similar objects or processes in the same era?
  - —How is the object appropriated/personalised?
  - —What range of social meanings is possible for this object and its relations? How 'fixed' might meaning/s be for this particular object?
    (Remember the screwdriver example please.)

The methodological strength in studying physical objects then (as I see it) lies in their value in stimulating creative questioning. Unlike diaries or films or interviews, physical objects and traces do not actively tell a story, at any rate, not a 'full' story. They are not 'totalising' (whole) in design or fashion. Rather, they are more often simply the *props* to a story which begs the question, 'What is the plot here?' Hence, the value of the study of objects is not exhausted by the reading of culture as a text. Rather, material culture can be read and re-read as props and prompts to many texts and many audiences.

Ethics in a material world

Generally speaking, ethical issues which focus on harm and consent are not obvious when studying wear on door handles or books. Reflecting on the nature of food in certain types of restaurants and similar activities is surely permissible for anyone and is harmless enough. Most observations in the material world are asocial affairs that require only a careful and sharp eye for detail and context.

But other examinations of material culture may not be so straightforward. Rathje, for example, in his studies of council tips has found human body parts and evidence of several other *crimes* such as theft and drug trafficking. What are a researcher's responsibilities here? You should be aware that studies which take you to garbage may lead you to inadvertently discover objects which were never intended to be seen. Tied in to this problem of crime and the dilemma of reporting it is the issue of *privacy*. Can garbage be regarded as 'non-private' once it appears on the nature strip in front of the house? Because the owners of the garbage can be identified this is an issue which needs thinking through.

And if you are employing friends or co-researchers in garbage studies, or toilet graffiti studies, there is also the issue of the *health and safety* of those whom you enlist to work with you. It really is an ethical concern to ensure that they are appropriately protected with gloves and clothing and that your team is organised to look out for each other. A co-worker who is accidentally folded into a hillside of garbage by earthmoving equipment is an event likely to have, among other things, negative consequences for morale in general and safety in particular!

Also, working in tips or toilets, or observing the floor layout of your local supermarket, may also require you to obtain *consent* from the owners or local authorities. This may be important for health and safety reasons or it may be so as not to arouse the suspicions of shop detectives. Observations of material objects or settings that require you to linger near children's playgrounds or schools can also arouse anxiety and suspicion. If not for ethical reasons, then for reasons of showing consideration towards others, it is a good idea to alert others to your presence and purpose. Most of the time, studying the material will not directly involve other people but remember that, often, *other people* will be observing *you*. They may not necessarily hold innocent interpretations of your activities. You need to reflect on this, take it into consideration, and if appropriate, inform the relevant people.

Finally, when examining physical objects—gravestones, books, ceramics, cutlery or other culture-specific items—leave them as you find them. Fight the tendency in every Westerner's breast to take souvenirs or memorabilia. In simple terms, taking small tokens of objects without the owners permission is *stealing*. This ethical issue might seem simple, obvious and needless to say but lots of people, from Egyptians to Australian Aborigines, have lost valuable objects to past researchers. Look and then leave. If you want physical reminders of the objects, photograph them!

Recommended reading

Bayley, S. (1986), *Sex, Drink and Fast Cars*, Faber & Faber, London

Deetz, J. (1977), *In Small Things Forgotten*, Anchor Press, New York

Gilbert, L. (1980), *A Grave Look at History*, John Ferguson, Sydney

Gould, R.A. and Schiffer, M.B. (1981), *Modern Material Culture; The Archaeology of Us*, Academic Press, New York

Hodder, I. (1989), 'This is not an article about material culture as text', *Journal of Anthropological Archaeology*, vol. 8, pp. 250–69

Klofas, J.M. and Cutshall, C.R. (1985), 'The social archaeology of a juvenile facility: unobtrusive methods in the study of institutional cultures', *Qualitative Sociology*, vol. 8, no. 4, pp. 368–87

Miller, D. (1988), 'Appropriating the State on the council estate', *Man*, vol. 23, pp. 353–72.

Vera, H. (1989), 'On Dutch windows', *Qualitative Sociology*, vol. 12, no. 2, pp. 215–34.

7

Simple observation

Most books written on observational methods, particularly in the field of anthropology, write about participant observation. In participant observation, the researcher interacts with the people that he or she is studying and makes observations in the course of these exchanges. The researcher records the day's events, social activities and the people met. This is done as part of other activities such as interviewing and using informants. This chapter concerns itself with simple observation—the observation of the unobtrusive researcher, the watching and listening of the detached onlooker.

Although there are many studies which employ observation as their primary method, there is a surprising lack of books or articles telling one how to actually achieve this technically. Because observational work is mainly understood as simply *looking*, the technical discussions have always been tempered by the qualification that 'it depends on what you choose to observe'. Schatzman and Strauss (1973) in a well-known book about naturalistic field work in sociology have several chapters dealing with strategy. However, these 'strategies' are not technical but rather social strategies. They are discussions about different roles the observer may assume—passive, active, hidden, limited and so on.

In this chapter, I will outline the types of simple observation that the unobtrusive researcher has at his or her disposal. I will then briefly review some studies which exemplify some of the key observational types. The review will begin with sociological and anthropological work but will also include some ethological (animal behaviour) studies and commentary. Ethologists have made sophisticated use of the observational method for obvious reasons and some have applied these skills to human behaviour, linking behaviour with biology. However controversial these types of links may be, ethologists have developed a slick and significant body of practical advice on how to observe. I will be summarising some of this useful material after a discussion of the advantages and disadvantages of the method in general.

Types of observation

Webb and his colleagues (1966) and Denzin (1970) who summarises them, mention the major types of observation.

Types of observation include:

- exterior physical signs
- expressive movement
- physical location
- language behaviour
- time duration

Exterior physical signs refer to observations such as clothing, graffiti, street signs, menus, tattoos, tribal markings, hairstyle, shoe style, jewellery, houses, calluses, radio transmissions, church services and commercial products. Denzin (1970) goes on to provide an example of a study of anti-Semitism in American communities which, among other things, noted the number and placement of swastikas on buildings, sidewalks, posters and billboards.

Expressive movements apply mainly to bodily movements such as eye, face, limbs, body posture. Also included here are observations of frowning, smiling to the extent of uncovering the teeth, gait, handwriting and gestures. Harper and Wiens (1979) argue that the study of expressive movements can demonstrate how these might either underline and support or contradict the verbal; fill out or repeat the verbal.

The observation of physical location includes not simply human use of physical settings and those settings themselves, but also the study of personal space. Many studies document the often observed relationship between physical closeness and quietness and the personal or confidential nature of conversations and relationships. The

The study of expressive movements can demonstrate how these might underline and support the verbal or fill out or repeat the verbal. Courtesy of the *Geelong Advertiser*

reverse is also true. Denzin (1970) believes that studies of interactional norms for lifts, air flights, train trips or even ocean liner voyages can yield valuable insights into what people do in confined spaces. Webb and his colleagues also discuss ethnic clustering behaviour or even the signalling behaviour of motorist's vehicles. They cite another study which shows that telling ghost stories to a circle of young boys at night will reduce its circumference in a short time.

Observing language behaviour includes examples such as noting stuttering behaviour, slips of the tongue and general conversational analysis. Of the latter, Webb and his colleagues describe the study of one researcher who walked up one long busy street at 7.30 p.m. every night for several weeks. The record of conversations overheard revealed that men talked about the opposite sex in about 8 per cent of conversations while women did the same in 44 per cent of the overheard conversations.

Time duration studies refer to the time paid to an exhibit, shop or different type of interaction, reasoning here that length of time indicates depth of interest. The observation of lifestyles of families or villages can fall into this category also, documenting when people come and go and perform one or another task at what times of the day or week.

Observations have the major advantage of being able to be conducted just about anywhere in public culture. Of course this means that private areas are not usually accessible without permission and without becoming in some ways, a participant observer. Nevertheless, when 250 undergraduates in a first year sociology class were asked to go out and do a small observational exercise, the variety of social situations observed was rich and diverse. The box on page 119 identifies some of the observed situations to demonstrate this.

The undergraduate exercise revealed a good many things about the way people conducted their first formal observations. However, the first thing one notices about the list is that they are observations of gross behaviour and social setting usually of the time duration type. Few people attempt, for example, observations of physical signs. And yet, a lot of fascinating work has been done here, particularly in the study of clothes. Let me illustrate.

Webb and his colleagues (1981) discuss how various studies have used the study of clothes to illuminate their findings about class, status, gender and social attitudes. They cite studies of shoe styles, the 'flashier' the shoe the more culture bound the individual.

Sample first year sociology observation projects	
pool parlour	child care centres
cinema	supermarkets
trams/trains	libraries
gallery/museum	restaurant/cafes
food court	schools
church	cars at traffic lights
football crowd	pedestrian crossing
college kitchens/laundry	pinball parlours
strolling couples	camping grounds
shops	bus rides
birthday parties	family gatherings
pubs/bars	shop entrances
malls	bus stops
TAB/betting shop	beaches
nightclubs	parks

They examine the relationship between attire and academic performance at university (the better dressed gained the better grades!). Also cited is a study of clothing style and petition signing behaviour. Neater dressers were more reluctant to sign implying that 'neat dressing' was some general indicator of social or political conservativeness.

Barthes (1985) conducted a semiotic analysis of women's fashion. His style of semiotics, however, sees clothes as a language system quite literally. He attempts to describe clothing as a sign system whose grammar, syntax and meaning are a formal language system of their own. Davis (1989) does not quite agree with Barthes literal interpretation of the 'language of clothes'. Clothes are a more indeterminate and ambiguous set of signs, much like the reading of music. Except perhaps the case of uniforms, the meaning of clothes is linked to the ever-changing notion of 'what is appropriate'.

Davis reviews quite a few historical and social theories about clothes and it is therefore a useful article to look at if you intend conducting observations on clothing. His particular interest was in the 'little black dress' and blue jeans. Both items of clothing were the attire of the poor but these gradually became incorporated into bourgeois fashion as items of modesty and elegance. Little black dresses were the usual attire of maids from the mid-nineteenth to mid-twentieth century, something they wore with white caps and apron. After the Great Depression of the 1920s the 'poor look' became daring and vogue, promoted by Chanel's advice to dress 'as plainly as your maid'.

Cotton material dyed blue, which was said to have come from 'de Nimes'(denim) in France and worn by dockworkers in Geneo (Genes), was manufactured for a wider US market in the mid-nineteenth century by a man named Morris Levi-Strauss (Davis 1989). Similar to the little black dress, its humble origins quickly associated it with leisure, casualness and the outdoor life. Popularity with successively different groups attached even richer and equally attractive associations. Jeans were worn primarily by artists in the 1930s, hoodlum bikers of the 1940s and 1950s, and new-left political activists and hippies of the 1960s. An amalgam of images from these groups—freedom, protest, simplicity, art and rock 'n' roll, rurality and romance—was symbolised by this one article of clothing. If wearers were not consciously attracted by these associations, then (according to Davis) many onlookers made these associations and many manufacturers promoted them. As a well-known Levis advertisement asks in respect of these images, 'Do you fit the Legend?'

Palca (1981) also looks at clothes as language, not literally as Barthès has done or sociologically as Davis has done but, rather, metaphorically. People speak with their clothes. There is not one language in clothes, but several. When we observe clothes we must remember to include hairstyle, jewellery, make-up and other body decorations. Clothes can act as slang (jeans, sneakers, caps), as adjectives (trimmings, accessories), or they may be expressive of local and foreign 'phrases' (such as an Oxford University educated Arab in suit and turban).

Palca argues that clothes make statements. What they say depends on the context or situation. The same clothes worn by different people will say different things. Weight, height, ethnicity, age and gender all contribute to the complex configuration of signs. Clothes language may be formal such as uniforms that convey membership and rank. Clothes can also lie—think of undercover police work or job interviews. The language of clothes can also be superstitious such as when people wear articles of clothing or charms which are 'lucky' or 'blessed'. Clothes and their symbolic significance can also be picked up and amplified in the media. Williamson (1986), for example, cites 'leg warmers' for women as the symbol of the 'new woman'. Tall, leggy women in their jeans or aerobic gear prancing about in their leg warmers are the symbols of the modern young career woman who is still soft and sexy on

Little black dresses were the usual attire of maids from the mid-nineteenth to mid-twentieth century. Courtesy of the *Geelong Advertiser*

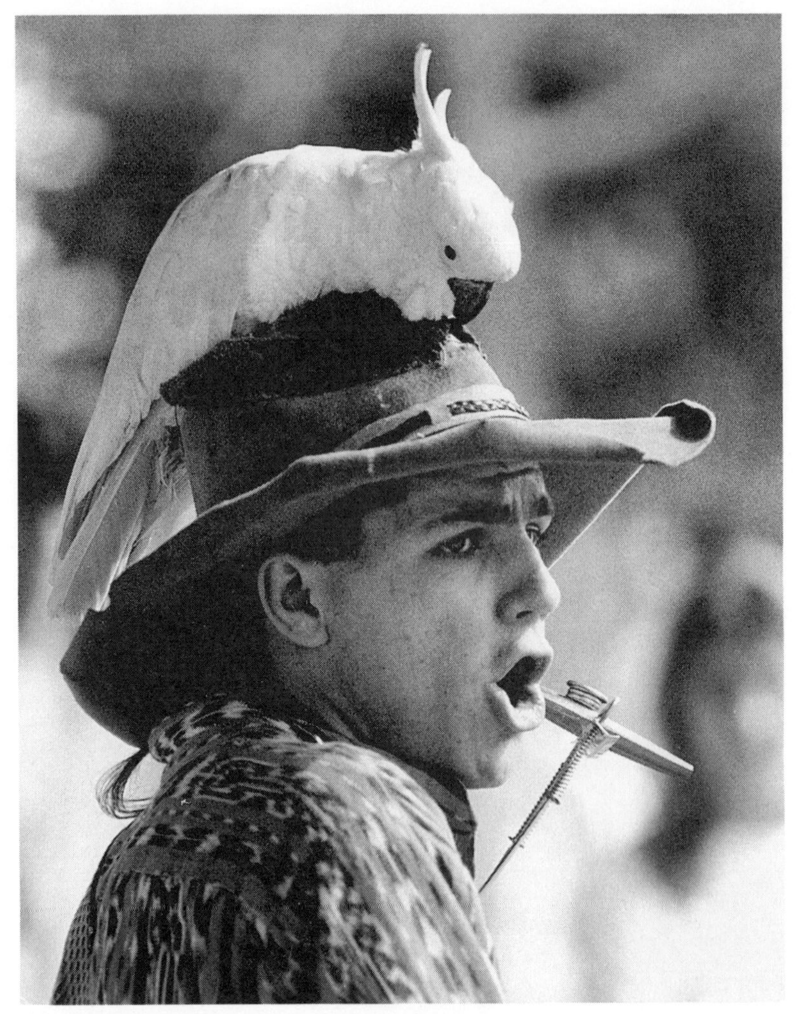

People speak with their clothes. Noting that this man is busking, and also noting his shirt, hat, hair and cockatoo, what is he 'saying' to those who see him? Courtesy of the *Geelong Advertiser*

the inside. These symbols are used to promote sexual pulp from short romance novels to magazines such as *New Woman* and/or *Cosmopolitan.*

Aside from exterior physical signs, a subject we discussed in greater depth in the chapter six, the observation of expressive movements has also not been popular with novices of observational method. Yet many of these types of studies can be quite simple to perform. Yule (1987) describes an observational study which she believes is so simple that anyone with a clipboard and some time can conduct.

Yule observed 85 adult–child interactions for three minutes each. These were matched for observations with adult–adult pairs. For each of the three minute observations she watched the facial expressions of the pair during their interaction and also noted other gross bodily movements. She concluded that many parents in her study were tough on their children: adults were rude, asocial, intolerant, insensitive and aggressive. Three quarters of the adult–child interactions showed little or no eye contact during talk by the adult. There was also a significant amount of physical pulling, shaking, slapping and verbal threats. Four-fifths of the adult–adult pairs, however, were courteous with much eye contact and smiling. Yule concluded that the child abuse, in this context, may be the extreme end of normal behaviour rather than an abrupt and different piece of deviant behaviour.

In a twist to analysing expressive movements in people, Nash (1989) looked at the expressive movements and physical features of English bulldogs. Knowing that people attribute 'social' meanings to the animals his study explored how dog expressions are classified and understood by people. Of course, interviewing the dog owners was the main method employed by Nash, but an essential part of the research was also based on simple observation of the dogs themselves, an interesting innovation. But dog owners are not the only serious observers of animals.

As I mentioned in the beginning of this chapter, ethology has applied its wealth of observational skill not only to animals but also to human activity. They have been particularly interested in expressive movements especially in studies of children and individuals with psychiatric conditions (Huntingford, 1984). Eibl-Eibesfeldt (1975) argues that ethology can also offer theoretical insights into behaviour drawn from biology. Now, biological explanations of behaviour have been unpopular since the 1960s when the social sciences strongly criticised any suggestion that biology *determines* behaviour, particularly in the gender area.

However, many ethologists (sociobiologists aside) do not make these kind of claims. Eibl-Eibesfeldt, for instance, argues that biological influences go beyond hair and eye colour. Ethology is not designed to be a reductionist theory of human behaviour but the influence of biology on behaviour suggests limits to cultural explanation. 'Innateness' does not equate with unalterable or unchanging. On the other hand, some biological hypotheses do seem to explain the inexplicability of certain examples of widespread behaviour. In children or infants for example:

- the rhythmic head searching for the nipple following a touch of the mouth
- hand-grasping reflex especially to hair
- 'automatic' swimming and walking movements in the newborn when placed in water or placed on a firm surface
- smiling and visual fixating (even in blind infants) and babbling (even in deaf born infants—this also occurs in children born without arms who cannot 'feel' smiles)
- laughter, anger and the fear of strangers

The strength of these in-born qualities can be assessed by examining this behaviour in different cultures. Cultures do not supply exceptions but rather different intervening influences on the innate. Sometimes this is a modification which leads to accentuation or deletion of the trait. Eibl-Eibesfeldt (1975) also suggests some adult examples.

- Photos taken from a height show people approaching conspicuous landmarks (for example, flagpoles) for no special reason
- Those who eat alone look up and 'into the distance' after each bite or two (something primates also do)
- People who are shown photographs of faces expressing various emotions are generally correct in selecting the emotion (this applies irrespective of whether the people are from Japan, USA, Papua New Guinea or Brazil).

I am describing some of Eibl-Eibesfeldt's observations for two reasons. First, asking people about their expressive movements may not always be satisfactory. They may not know or they may give explanations which are not insightful causatively but instead are rationalisations. Werner and Schoepfle (1987) in their massive work on *Systematic Fieldwork* in anthropology suggest that all ethnographic 'observations *must* be submitted to comments by the natives' (p. 260). This is fine to assess the emical meanings if there are some (not everyone has an explanation for what they do). It is also probably very useful in understanding how different people rationalise or explain the same behaviour in different social settings and in different cultures. (For a good example of this point see Geertz's discussion of winking behaviour: Geertz, 1973, pp. 6–7). But the emic may not be the main reason behind the behaviour. This brings me to the second issue.

There is little point in using the emical meanings or your own cultural theory (etic) to explain why a behaviour occurs if there is a biological reason for its existence. Do not misunderstand me here. How people come to understand their behaviour is important (emic). How we come to understand their understanding of their behaviour (etic) is also important. They may both be useful insights into the world in different ways. But an understanding of behaviour may not be exhausted by placing ticks beside the categories of emic and etic. Furthermore, the cultural construction of meaning does not, in every case, explain why the behaviour itself is present. Some cultural constructions are responsible for the presence of widespread examples of behaviour. Some do not. In these cases, cultural constructions may be rationalising devices. For observations which are not informed by emic inputs, especially observations of children, a basic familiarity with ethological writings can come in handy—if only to contextualise your own theory for the behaviour. The humanities and social sciences do not have a monopoly on useful theories of culture and human behaviour—no matter what they tell you in sociology or anthropology.

Some problems in observational work

Let me begin this section with the good news. The real strength in the observational method is that it forces you to become familiar and objective with the subject of your enquiry. If 'objective' is too ambitious or positivist a word then perhaps the idea of familiarity will do, getting better acquainted with the subject of your enquiry. In fact, closely observing something or someone can give one a new perspective, a perspective different to the usually less interested gaze of ordinary observation. In this sense, 'the simple observations of the detached observer' are a series of misnomers. Simple observations are anything but simple, because they are highly *systematised* and thorough and more intensive ways of looking and listening than usual. And the observer, far from being 'detached', is, in fact, thoroughly engaged, indeed, engrossed in his or her subject. This is just one of the real values in systematic observational work.

This intense examination suggests a second advantage. Features which may be overlooked casually are given greater opportunity to be identified by the observer. Huntingford, (1985) argues that the attention to detail and sequence required in any systematic observation can transform the pedestrian into the intriguing. Third, Denzin (1970) reminds us that what people actually *do* is often more important than what they say. Remember the study of screwdrivers in chapter six? Finally, observations are wonderfully unobtrusive and even when they accidentally become noticeable, the observer-effect wears off in time (Berry, 1979). In public places, and particularly in busy places, a casual air, minding your own business and a pair of sunglasses can enable hours of quality observation. Staring behaviour, dressing as a plain clothes police officer or shop detective and furious note-taking will ensure a short career or trial of this method.

Advantages of the observational method that:

- it forces the observer to familiarise with the subject
- it allows previously unnoticed or ignored aspects to be seen
- people's actions are probably more telling than their verbal accounts and observing these are valuable
- it is unobtrusive and when obtrusive, the effect wears off in reasonable time

The first major disadvantage with simple observation reminds me of a famous line from the screen actress Mae West, 'Is that a gun in your pocket or are you just glad to see me?' As Ms West's line ably demonstrates, when we see unfamiliar objects and behaviours we tend to subsume their function or meaning into those that we are familiar with. This is a life-long habit. This type of observer bias can attract attention to the unusual and take for granted the familiar. In turn, the unusual can be recorded in loaded or ethnocentric ways. Other sources of observer bias can be fatigue or colour blindness in the observer, boredom or undisciplined/unsystematic attention (Denzin 1970). Observations can also be time consuming in the field, and when recording these, can generate large volumes of data which can be difficult to manage and analyse. Complex or busy patterns of interaction can be confusing particularly for the single observer and this may lead to attending to minor examples of behaviour or events. This can lead the observer to miss 'the forest for the trees'.

If observation is used as the only method, the usual emic/etic problems come into play. Observation is quite a good method for small populations but becomes difficult in large populations and, as mentioned earlier, tends to be confined to public areas of social life. If recording the observations does not occur immediately, recall can be a problem.

The High Court has called eyewitness testimony "notoriously unreliable".

We have brought you stories of a witness identifying a barrister as a wrongdoer only to be followed into the box by one who claimed the solicitor at the bar table was the real crook.

• Last December we told you the outrageous tale of John Watkins who spent 11 months in jail after allegedly confessing to a bank robbery he didn't commit.

Six eyewitnesses incorrectly selected the innocent man's photograph as that of the bandit who held them up. While he was in jail the real villain struct again and again.

• Hit man Chris Flannery, missing but not missed, was himself the subject of murder attempts.

His wife Kath blamed Tom Domican.

She claimed to have seen him jogging near her home. It was not him. Domican had a broken leg in plaster at the time.

• In a famous US case, Robert Dillen, a freelance photographer was arrested for urinating in a park.

He denied it but paid a small fine to avoid court.

His photo in the police files made his life hell.

The next day it was picked from a batch of 10 by an armed robbery victim.

He was charged and released.

Over the next few months he was charged with three more armed robberies and a kidnap/rape.

His wife and career were gone by the time the real culprit was caught and confessed.

Research shows that when a witness sees a photograph it seems to over-ride or even replace the memory of the original face.

Experiments suggest that the new image supplants the old.

Test subjects who pick a wrong photo have been shown to stick with the photo subject in preference to the real ciminal in later line-ups.

Whatever the reason, mistakes abound.

If recording the observations does not occur immediately, recall can be a problem. From 'Murphy's Law' by Chris Murphy, *The Sun-Herald*, 4 October, 1992. p. 11.

Even when recording on the spot, accurate record keeping can be a difficult task without experience or strong pre-field planning. Finally, Denzin (1970) and Martin and Bateson (1986) warn of population instabilities. It can be difficult to watch some subjects who come and go and disappear for lengthy periods. One's sample of time and place may not be representative of the character of that place or series of events.

These problems raise issues of reliability and validity as well as

Methodological problems to look out for are:

- observer bias (1)—selectivity, the unusual attracts, the familiar taken for granted
- observer bias (2)—fatigue, colour blindness, boredom
- inadequate or biased recording
- the generation of large amounts of data
- complex or busy patterns which can confuse
- emic/etic problems if used alone
- limited usefulness in large populations
- limited usefulness in public places
- population instability
  —difficulty of keeping tabs on chosen subjects
  —different environments contain different people at different times
- recall problems if recording not immediate

emic and etic. Observer bias can be checked using inter-rater reliability practice—use more than one observer and compare each other's categories and observations. This use of more than one observer is a particularly important piece of advice if the environment is busy or complex. There are many ways to tackle the problem of observer bias. One way is to use cameras or videos when recording the observations so that the data can be re-checked and cross-checked by others. Another way is to simply be aware that this will be a problem and to monitor boredom, taking frequent breaks, and ensuring that observation periods are not too arduous or taxing. Good recording practice, with experience, will encourage faith, yours and others, in your records. We will discuss the options in a moment. Recall problems can be dealt with by minimising the time between observation and its recording. Again though, sometimes the way one records may help overcome this, that is, the use of cameras of one sort or another.

It certainly is a problem occasionally to keep an eye on everyone or the one person you wish to observe when they are obscured by crowds, go to the toilet to buy drugs, disappear up and down lifts or turn their backs to you. Unfortunately, not all problems have an answer and this may well be one of those. Ensuring that you observe behaviour in contexts that reveal important or typical features is a sampling issue. Clear observations made at certain times and places will have implications concerning representativeness due to observations made at those times or those places. These must be considered, as with many of these problems, in the planning stage prior to the field observations.

Unfortunately, as with many other methods (such as archival work) the methodological problems are not the only ones. There are also practical problems. Martin and Bateson's (1986) comments about animal observations are equally instructive to observations with humans. They warn of poor visibility, bad weather, 'complexities of habitat', isolation and physical danger (p. 100). The undergraduates who attempted their first observations for an assignment also recollected their practical problem 'in the field' and these reflect Martin and Bateson's warnings in dramatic and telling ways. Students complained of:

- assault/hostility—one person was beaten up in a snooker parlour (remember not to stare!)
- speed of events made recording extremely difficult
- recording was difficult because many chose a prose style of documenting events
- people staring, curious children (sunglasses here please)
- security guards ('Why are you lingering about here—move on')
- curious waiters/waitresses ('What are you writing underneath the table?')
- chosen field area too large to watch (pick another?)
- drinking or eating too much to look 'normal'
- 'pick up' attempts, especially for female observers
- hostile shop-keepers ('Why hang around if you're not going to buy?')
- poor weather

As you can see from a perusal of these problems, quite a few relate to unsuccessful attempts at simple observation. Many people looked like they were looking. In our culture, people who look a lot are rarely interpreted as researchers. Rather they are signalling something sexy, menacing or curious. This begs an approach or comment by others. Berry (1979) suggests sunglasses for those who cannot help staring and who work in daylight. Other people overdo their attempts to look normal and either leave the field bloated with food or drunk. No-one except you counts how many bites of a meal or slurps of a drink you take so 'steady as you go' is advisable. Other problems are practicalities which again, experience should overcome.

Site selection and record keeping relate to the technical practices of this method that we will now turn our attention to in some detail.

Site and sampling

Irrespective of the type of observations that you plan to make, there are a few areas of observation that you should either note or at least be aware of as Runcie (1980) explains:

1 Note the physical characteristics of your chosen place. Werner and Schoepfle (1987) divide the setting into three aspects: the location (Sydney), the stage (a jewellery shop) and the set (the counter or the backroom etc.). Sometimes a drawing or some photographs of the setting are useful for later reference and explanation.
2 Note the actors, what they do in terms of gross and finer movements. Also note what they look like (age, sex, clothing, etc.).
3 Note behaviour cycles, the duration and frequency of certain acts.
4 Note the stage/period/phase in which this behaviour occurs within the setting (lunchtime at university, for example).
5 Note the state/period/phase in which the setting itself is placed (term time at university).

In summary then, one should note place, objects, activities and contexts of the setting. If one is interested in the physical space then the other areas will be less important. If one is interested in behaviour then the bulk of your notes will concentrate on these. Nevertheless, notes should exist about the other areas to contextualise the subjects and the observations themselves.

There are four main ways to sample when making observations. According to Martin and Bateson (1986) these are ad libitum, focal, scan and behaviour sampling. Ad libitum sampling is impressionistic and non-systematic. The observer simply records whatever is of interest. This is particularly useful if any observer feels that a particular set of behaviours is significant or revealing given the setting. A one-off tantrum by a person at a cafeteria queue is a good example—a rare piece of behaviour that may be more revealing socially because of its rule-breaking significance. Ad libitum observations are capable of capturing these types of performances because most social detail is ignored until something of interest arises. In this sense, ad libitum sampling is also useful for pilot or preliminary observations as a way of deciding what further systematic work needs to be done. As a sampling technique used alone, however, bias and reliability can be problems.

Focal sampling involves choosing an individual or group of persons and recording all their behaviour/physical features over a specific period of time. Yule's study of adult–child pairs over a three minute period is a good example of focal sampling.

Scan sampling involves rapidly scanning a whole group or individual at regular intervals and then recording this. Usually the recording is simple such as noting the presence or absence of certain activities. A single scan can take seconds or minutes, but it usually takes in the whole physical scene and the individuals in it.

Behaviour sampling involves simply choosing a behaviour and noting who does it and when it is displayed. You might choose to note smiling or touching movements in a group and note who in the group does this and the context which prompts it. Ad libitum and scan sampling are sometimes known by the more general title of descriptive sampling while behaviour sampling is sometimes called selective sampling (Werner and Schoepfle, 1987).

After choosing your sample you will need to decide whether you will record continuously or intermittently. Continuous recording can be quite arduous, particularly with ad libitum samples. Intermittent sampling is easier but has other drawbacks. If one records intermittently, there are usually three ways to do it: either one records the observation every ten or so minutes and records the behaviour *at that time* (instantaneous record), or one records the behaviour which occurred in the whole preceding ten minute period (one zero sampling) (Martin and Bateson, 1986, p. 57). One zero sampling can lead to over and underestimation of activity because one may assume from the record that an activity has occurred throughout the timeframe. On the other hand, one might assume that a behaviour occurred once when in fact it occurred several times. It is important to note frequency in this context so as to clarify this issue when later reading over the record. Instantaneous and zero one sampling tend to be favoured among ethologists and psychologists. The third type of recording, periodic recording, is common practice among ethnographic fieldworkers. These people are particularly concerned that the actual act of recording does not attract attention. In this context the record is made at the end of a time period, decided before entering the field, say, every hour or at the end of the day. The recording takes place regularly and intermittently but the time periods are arbitrarily decided usually on the basis of convenience or some judgement by the ethnographer as to what and when it is most important to record. Obviously one major problem with periodic sampling is imperfect recall and recording.

Main sampling techniques are:

- ad libitum—impressionistic
- focal—group/individual over time
- scan—group/individual intermittently
- behaviour—note when behaviour occurs

The recording of these samples are conducted:

- continuously— everything recorded in an ongoing manner
- intermittently
  - —instantaneously
  - —one zero sampling (recording periods retrospectively)
  - —periodic

Recording devices and styles

The best advice about recording devices and styles that I have been able to find comes from Brandt (1972) and Martin and Bateson (1986): to record your observations you really have about five main choices—the camera/video record, the tape recorder, written notes, checklists and ratings.

Camera and video are particularly good for cross-checking and re-analysis. If you need to re-code, photographs and film footage can make this easier too. Still or moving images are great records if the setting you are observing is complex and/or fast moving. Looking over the images repeatedly allows you to leisurely go over the material for aspects of the setting that you were not able to notice on first inspection.

The tape recorder is valuable for doing for sound what the camera and video do for visuals. Recording the sounds of events or behaviours or settings can add another dimension to any analysis, allow the visual texts to be compared and question the sound texts and vice versa. Additionally, tape recorders can also be used to dictate field notes. Instead of actually writing one can speak one's thoughts and observations onto tape. Of course, eventually you may want these in written form and transcription will add considerable time to later coding and analysis of the observations.

Written notes are the traditional way of recording observations, usually in short- or longhand. It is best to do a bit of planning about this before entering the field. Organise some codes or abbreviations *before* you enter the field to make the recording task lighter and so that writing time does not erode observation time. According to Brandt (1972, pp. 80–94) observational notes can take four main styles. They may be: anecdotal (these are often descriptions of critical incidents and are useful in ad libitum sampling); specimen records (where a place and person/s are chosen and the recording continues over time); field notes (these place less emphasis on strict description—some analysis as this occurs to you in your observation is also recorded, perhaps in the margins of your notes, and so conceptual material becomes dotted through the descriptive observations); Werner and Schoepfle (1987) also argue that a good field journal should contain self-observations, a record of your own reactions and feelings about what you saw and heard. This might be done at the end of a day of fieldwork, back in the home, tent or cave where one is staying. Finally, observational notes might also include ecological description. These are details of the environment in which behaviour is taking place, for example the spatial organisation, the relationship between buildings, objects, rooms or traces. Note dominant colours, shapes, styles, frequency and materials of various objects in the setting. These can later be linked with other research on housing or clothing fashion or colour psychology (Luscher, 1971) or technological genre.

Written descriptions should be recorded as soon as possible if these are not to take place during the observation. Try to write as much detail as possible, quote talk if relevant and feasible. In any case, with recording of behaviour, be sure to preserve the sequence. When recording in the observational setting itself, Runcie (1980) warns of being as discrete as possible lest you attract unwanted attention. Be technical in your description and avoid unreliable descriptors such as 'well dressed' and 'dirty jeans'.

Checklists are another recording device which can be useful. Checklists are usually grid-like charts using columns. On one side the behaviour categories are listed and on the other side the time intervals. Martin and Bateson (1986) also recommend a remarks column for the unexpected observation or comment. Obviously checklists come into their own in scan and behaviour sampling. See example on page 134.

Observation (scan) of six couples in Blue Moon Cafe
Monday 7.30–9.30 pm

Time interval	*Touching*	*Dancing*	*Hanging from the rafters*	*Remarks*
7.45	(3)	(1)		
8.00	(2)	(1)		
8.15	(4)			
8.30	(1)			
8.45	(1)	(2)		The drunk couple
9.00		(3)	(1)	hanging from the
9.15		(1)		rafters were the
9.30				bartender and waitress

Checklists can contain static descriptors such as age, sex, time-intervals and so on and action descriptors such as touching, dancing and hanging from the rafters. However, do not forget that static descriptors can be paired with other static descriptors such as clothes types, jewellery wearer or not, hats/caps, left- or right-hander and so on.

Brandt (1972) also mentions other types of checklists which may be useful. The activity log, for example, which starts and finishes at set times and notes the major action or events. The discrete event log which records frequency of events or features (for example, those buried before 1900, those after). Performance records such as score sheets (golf/tennis scores, the number of televisions that a person sells, for example) and the trait checklist. This trait checklist records instances of anger or humour or classroom attention etc.

Finally, ratings can be used to record observations. Sometimes ratings use numbers, at other times they rate graphically. Usually one quality (active, fast moving, constantly talking) is placed at one end and its opposite (passive, not moving, silent) at the other. In between these opposite poles is a sliding scale of qualities (see example on next page).

Sometimes the items on various checklists are given a score and then these are added to give an overall score which indicates level of activity or trait (cumulative rating). Other ratings are simple forced choice ratings, often checking to see if a quality or event is present or not. Reliability has been a bit of a problem with ratings but when other observers compare their ratings and the categories are clearly defined and unambiguous this can be controlled considerably.

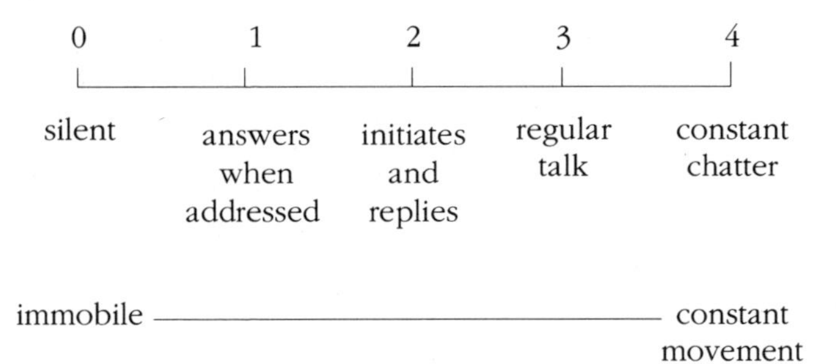

Present	Absent	Similar behaviour (state)

Available recording devices are:

- camera and video recorders
- tape recorders
- written notes
  - —anecdotal
  - —specimen
  - —field notes
  - —ecological description
- checklists
- ratings

Observing ethics

At first it might appear strange to talk about the ethics of observation but then, one person's looking is another person's spying. Some people believe that researchers should ask permission before conducting observational work but of course there are two other problems that this would create. First, if people know that they are being observed this can alter their behaviour (reactivity). This is a particularly important point for short observational cycles such as those exemplified by Yule's (1987) three minute observation of child–parent interactions. For longer observation periods when observer effect may wear off, consent might be feasible. However, this brings me to the second point. With large groups (football crowds) or social situations of high turnover (public bars) who does one approach for consent? If one is observing children in a school playground over several hours or days it is wise to gain the consent, or certainly to at least inform, the school authorities. Strangers lurking around school yards are always the understandable focus of suspicion and fear from teachers, parents, police and the children themselves. In this context, consent operates as a way of minimising the harmful effects of such anxiety on all the people concerned.

On the other hand, observations taken at football stadiums and university campuses are recognised as public places. Even if one requests permission from the relevant authorities to conduct the work there, it is nigh impossible or impractical to seek permission from most of the many people one might observe.

Another issue hotly debated by ethicists concerned over social sciences observation is the topic of privacy. People expect, and, indeed, have a certain right to, privacy. But it is doubtful that privacy is invaded by being observed in a public place. On the other hand, the observation of private acts in public places—such as using a public toilet—may be construed as an invasion of privacy. This is a delicate and complex matter but the debate of the different and subtle issues is well rehearsed by Laud Humphrey's (1975) work *Tearoom Trade: Impersonal Sex in Public Places.* Laud Humphrey's study was an examination of homosexual acts in park toilets. Although interviews were conducted as part of the study, the initial work was observational. The study raised a furore in the United States on ethical grounds and the essence of that debate is captured in a series of critical articles with an accompanying rejoinder in the book's postscript (Humphrey, 1975, pp. 167–232). This is well worth reading, particularly if you have doubts about the ethical soundness of your chosen observational study.

Steps to simple observation

The preceding summary and discussion of ethological and anthropological advice should caution anyone from thinking that observational work is simply 'looking about and around yourself'. Observing is difficult. Recording is difficult. Planning is essential and a critical part of this is organising yourself carefully through the steps discussed in this chapter. Even so, practical problems will be encountered depending on the nature of the chosen problem to be investigated.

Except perhaps in ad libitum sampling, observational work should be carefully systematic—not only for the sake of reliability and validity but also to prevent confusion. Even in ad libitum sampling the actual recording and analysis should be carefully and systematically performed. I conclude this chapter with a step-by-step schema which might guide you when setting out to observe.

1 Define your interest:
  — an individual or group
  — a certain behaviour (language or expressive movement)
  — a certain setting (physical or social)
  — a certain class of object
2 Select your sample:
  — ad libitum
  — focal
  — scan
  — behavioural
3 Decide on your approach to recording:
  — continuous
  — intermittent: instantaneous/zero one/periodic
4 Choose one or several recording devices:
  — camera/video
  — tape recorder
  — written notes: anecdotal/specimen/field/ecological
  — checklists
  — ratings
5 Plan ahead:
  — familiarise yourself with the setting beforehand
  — draw or photograph the setting
  — perform a 'trial' observation to anticipate problems
  — develop shorthand codes, abbreviations, etc. for written notes
  — procure equipment
  — develop and test scales and ratings
6 Observe (emphasis varies depending on defined interest):
  — setting, i.e. physical, geographical and spaces and objects
  — actors, their appearance, movements
  — behaviour, solitary and interpersonal cycles
  — context (1), time and social occasion
  — context (2), historical and cultural
7 Analyse the resultant record through:
  — content analysis
  — thematic analysis
  — semiotic analysis
8 Write up final report (see chapter two)

Recommended reading

Brandt, R.M. (1972), *Studying Behaviour in Natural Settings*, Holt, Rinehardt and Winston, New York

Davis, F. (1989), 'Of maids' uniforms and blue jeans: the drama of status ambivalences in clothing and fashion', *Qualitative Sociology* vol. 2, pp. 337–55

Geertz, C. (1973), *The Interpretation of Cultures*, Basic Books, New York (see especially chapter 1)

Gregory, C.A. and Altman, J.C. (1989), *Observing the Economy*, Routledge, London. A good text in the ethnographic tradition. Chapter 4 has a particularly good discussion of the use of maps, aerial photographs and satellite information for observations in large areas

Gross, D. (1984), 'Time allocation: A tool for the study of cultural behaviour, *Annual Review of Anthropology*, vol. 13, pp. 519–58

Humphreys, L. (1975), *Tearoom Trade: Impersonal Sex in Public Places*, (enlarged edition with a retrospect on ethical issues), Aldine De Gruter, New York (see especially the postscript, pp. 167–232)

Martin, P. and Bateson, P. (1986), *Measuring Behaviour: An Introductory Guide*, Cambridge University Press, Cambridge

Miner, H. (1973), 'Body ritual among the Nacirema', in D.R. MacQueen (ed.), *Understanding Sociology through Research*, Addison-Wesley, Reading, Mass., pp. 14–18.

Runcie, J.F. (1980), *Experiencing Social Research*, Dorsey Press, Illinois, chapters 3–6.

Yule, V. (1987), 'Observing adult–child interaction—an example of a piece of research anyone could do', in M. O'Connell (ed.), *New Introductory Reader in Sociology*, Nelson, Edinburgh, pp. 69–75

8

Hardware and software

In addition to working in libraries and archives, and making observations of objects or people, there are also a number of useful technological aids for the unobtrusive researcher. Webb and his colleagues (1966) originally referred to these as 'hardware'. They concentrated on novel examples such as the polygraph, one-way mirrors, pressure gauges, electric eyes and electromagnetic movement meters. They also briefly mentioned cameras, microphones and audiotapes. Most people, however, do not have laboratories with one-way mirrors and many others would not know what to do with a 'pressure gauge'. This chapter focuses discussion on the technologies that most people are likely to own or have reasonable access to. These are the camera, especially the still and video camera, and computers. Integral to any discussion of these types of equipment are other issues such as film and programming. In that respect then, we will be discussing both hardware and software.

Because the issues surrounding photography have largely been discussed in chapters five and seven, the aim in this chapter is to provide a quick overview of equipment choices in photography and to sketch the methodological implications of those choices. In other words, I will attempt to alert you to what research techniques are possible or unlikely depending on your choice of equipment. As is usual, I will also discuss advantages and disadvantages of the camera as hardware and also note some ethical issues. The final section of the chapter will introduce readers to computer simulation and modelling. Many people either own a computer or have access to one and this is, and will continue to be, an interesting avenue of unobtrusive research, particularly for the computer literate.

There are four main advantages to employing some kind of hardware as an aid to unobtrusive research. First, use of hardware provides permanency of record. This first advantage is of critical importance because it provides researchers with the other three advantages which are: allowing that record to be amenable to reliability checks; providing an opportunity for theory generation and testing (this applies not simply to the initial theories used but also to other ones that may be generated later); being a useful aid to memory, particularly for complex data.

The general advantages of using hardware are:

- the permanence of the record
- reliability checks
- theory generation and testing, both old and new
- it aids memory

The main disadvantages are, first, that some hardware can attract attention either because of noise or the bulky nature of the equipment. Second, some equipment can be complex to operate and this can increase researcher time, error and bias. Third, some equipment can be expensive in its initial and also ongoing costs. Finally, nearly all equipment, of course, is dependent on the competence and general orientation of the operator. Therefore, the problems of hardware as method are often the problems of observation and interpretation in one form or another.

The general disadvantages of using hardware are:

- it can attract unwanted attention (especially cameras)
- it can be complex to operate, increasing error and time wastage (especially computers)
- it can be expensive
- it is ultimately operator controlled and dependent

Of course, each example of hardware has specific strengths and problems and I will address these particulars after a general discussion of the camera and computer.

The camera

Photography can produce photographs and film material which may be analysed in its own right as an insiders (emic) view of their own world and activities. The researcher may not necessarily be the one in control of the camera in this instance. This material may also be used as a basis for unstructured or semi-structured interviewing or it may be used as a means for constructing a photographic essay. But beyond one's imagination and native or trained ability to perform either of these tasks, the choice of equipment will also be a determining factor.

The use of the camera as hardware in unobtrusive research must not be seen as a consideration confined to the camera as machine by itself. True, there are many different cameras and this can make choosing one appear difficult. Nevertheless, in addition to the camera comes consideration of the important supporting hardware—lens, film, tripods, developing and exhibiting equipment to name only the major ones.

If you are a student financially surviving on part-time work and family gratuities, then my advice is to use the camera closest to hand, and get to know its limitations. However, if you are a postgraduate student or paid researcher, I suggest that you apply for some financial assistance. An expensive Hasselblad is not necessary for good research work, but neither should it be thought that one can easily get by with grand-dad's box brownie. Equipment costs money—the first question is how to buy the basics? Good equipment does not equal good research so the second question becomes what are the basics? The equipment jungle in the world of photography goes something like this (imagine this as a decision tree). There are three types of cameras: automatics, semi-automatics and fully manual operations. Fully automatic cameras are 'point and shoot' cameras with shutter speed and aperture settings fully automated. They are particularly useful for candid, spontaneous photography. In manual cameras, shutter and aperture settings are adjusted manually by the operator with the help of a (usually inbuilt) light meter. Some advantages of a manual facility of course are that one can under- or over-expose images; one can slow shutter speed to convey movement in 'blurred' action-type shots or 'freeze' very fast action; and one can take shots in very low light without a flash (by setting the shutter to open for a longer period). The main problems are really of two kinds. Manual cameras tend to cost more and, secondly, you need to know how to operate one.

Assuming that you can afford the few hundred dollars to buy one, learning to use a manual camera can be a fairly quick process, really only taking an hour or two of your concentration and perhaps a few rolls of practice film. Semi-automatics, of course, cater for all options.

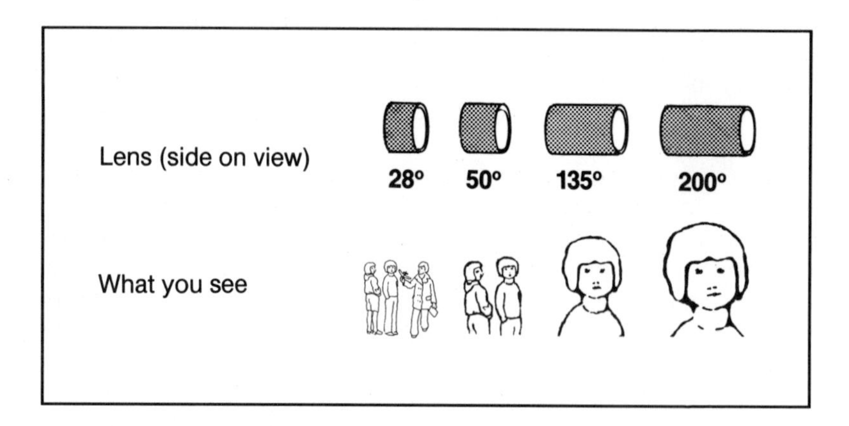

Whether you opt for a manual or automatic camera there is still another aspect to consider—whether to choose a fixed or interchangeable lens camera. Fixed lens cameras come with only one lens (usually 35mm to 50mm) which cannot be removed. Interchangeable lens cameras come with one lens (usually 50mm) but others can be bought to interchange with it. The disadvantage with the fixed lens camera is that you must do what the lens cannot. So, for example, if you wish to photograph close up, you must *physically* be close up. Conversely, if you wish to photograph a small room and its contents you may need to make many more photos than if you had a wide angle lens which could artificially 'miniaturise' the room for you. Ideally, you should select a single lens reflex (SLR) camera. This is because what you see through the viewfinder is what you will photograph. Non-SLR cameras contain viewfinders which do not look through the camera lens itself. This means that what you see is not necessarily what you get. The SLR should have some manual functions and lens changing facilities because this will allow most flexibility in research conditions.

There are also different choices of film: black and white or colour, print or slide film, fast or slow film. Practice and experience will help you to determine whether monochrome or colour film is most appropriate. Black and white film is a little cheaper to buy,

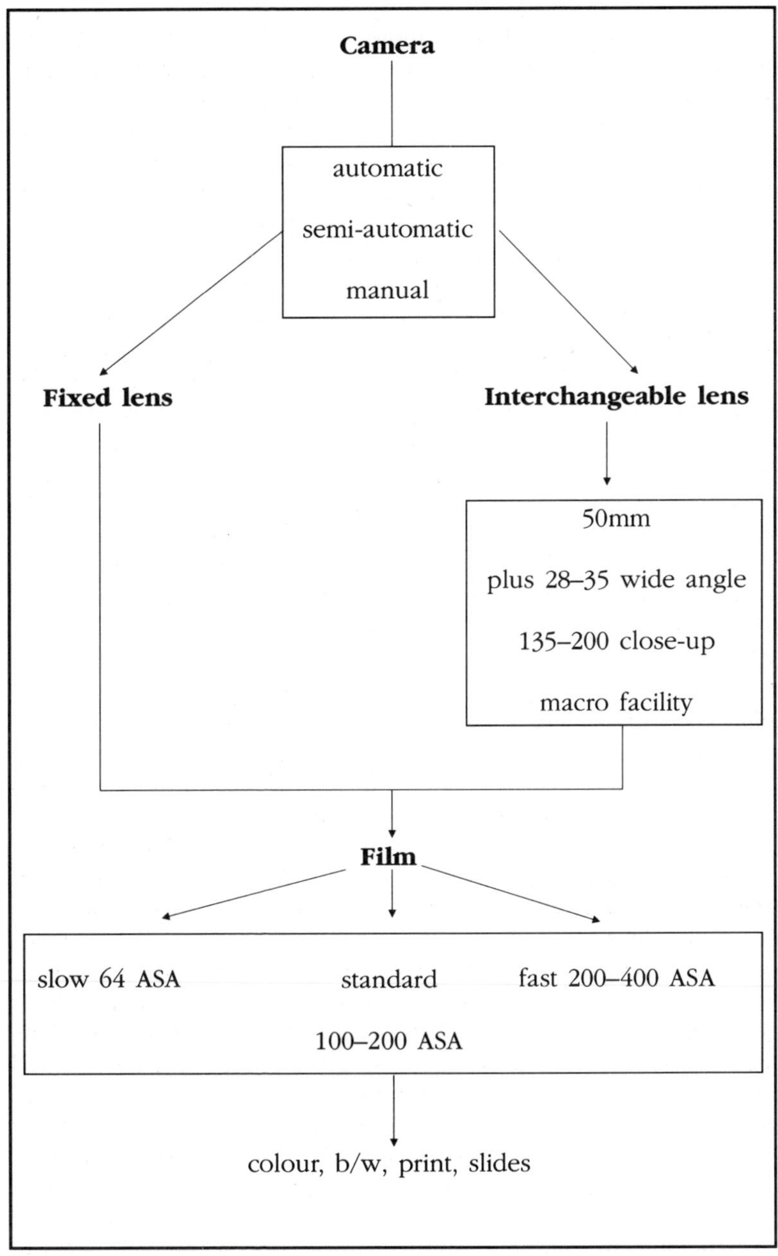

Your basic equipment choices for photography should go something like this.

but less widely available and more expensive to develop. Colour film is widely available but can become expensive to develop, particularly if you use a lot. Slides cost less but take longer to process and need a slide projector to view them. Hand-held viewers can be useful, but with hundreds of slides one can develop a tiresome squint. Film speed choice depends on what you wish to photograph and under what conditions: 100–200 ASA is probably ideal for most photographs in most conditions. Higher speed film such as 400 ASA tends to produce 'grainier' film, that is, images that are less sharp and clear. Low-speed films such as 64 ASA have a finer grain and tend to need good light, a flash, a good steady hand or a tripod. If you have 'shaky' hands or fingers which are attracted to lens surfaces, I suggest that you use tripods but, of course, this makes spontaneous photography very difficult. If you need to be free and spontaneous but still have the shakes try faster film, 200–400 ASA if automatic or, if you have a manual camera, do not shoot below 1/250 of a second.

On the question of lenses, the basic ones are as follows: 28–35mm are wide angle, 50mm is standard, 105–135mm are good portrait lens and anything above 180–200mm are close-up lenses. Most of the time the 50mm will be used but in photographing the inside of a houses, say, the 28–35mm range can be very useful. For expressive movement studies perhaps the 105–135mm range will come in handy—otherwise using a 50mm lens will require pointing the camera into people's faces far too closely. If you wish to photograph small objects such as jewellery items, small bits of litter or physical traces you will need a macro lens. Sometimes these are part of some models of automatic camera. At other times 'extension rings' can be bought cheaply for attachment to a standard lens for this super close-up requirement. If photographing people's activities cannot be performed unobtrusively without their taking notice or being self-conscious there is another lens which may help. This is given some mention by Eibl-Eibesfeldt (1975, p. 460) but is a lens well known in photography circles. A prism lens can be attached allowing the camera to photograph sideways. To the untrained observer the camera points away from them, but in actuality a mirror in the lens itself enables shots to be taken from the side of the lens.

Another gadget which may or may not be useful is a flash unit for photography in low light. I must point out, though, that it will be difficult to remain an unobtrusive researcher when you regularly flash your presence to others in this way. There are also accessories called motor drive units. These are attached to the base of the camera and allow multiple shots (say six per second) to be taken for as long as your finger depresses the shutter button (or when the film runs out, whichever comes first). Once again, motor drives can be noisy as any professional golfer will tell you. Nevertheless they can be particularly useful in the study of expressive movements, especially in capturing subtle sequences of detail. A long lens can give you distance and hence some way of correcting for the noise of the drive.

The video camera

The video camera is another piece of equipment that is essentially very similar to the still camera, except, of course, that it records moving images. The strengths and weaknesses of using both the still and video camera will be discussed at the end of this section. The choice of video camera equipment goes something like this.

There are standard VHS video cameras which use half-inch tape in video cassettes just like the ones you hire from a video rental movie hire shop. These cameras can fit into a small suitcase. Much smaller than the VHS video camera is the non-standard 8mm video camera. This is a physically compact piece which is easily hand held. The problem with 8mm video cameras is that they are considerably more expensive and they also need a special 8mm playback machine. However, being so small they may be more attractive for unobtrusive research purposes. Between the larger standard and quite small 8mm non-standard video camera is the compact VHS video camera. This is considerably smaller than the standard VHS video camera (less than half its size) and uses much smaller videotape cassettes. The compact video can also satisfy the need for physically smaller, lighter weight equipment but it has the added advantage that the video tapes can be used on standard VHS playback equipment. This is because its smaller video tape housing can be clipped into an adaptor for standard playback.

Popularity in private homes and academic institutions indicates that standard (including compact) VHS video cameras are the ones most available. Nearly all of these are fully automatic now and the better quality ones have a manual override which *is* important methodologically since, for example, auto focus takes too much control away from the operator. If you wish to pan from background to foreground (or vice versa) manual focus ability can make this much easier. Similarly, control over the aperture (called iris in VC language) can give you greater control over brightness of picture, and night and day shots.

Unlike a standard still camera, the video tape is not actually film but rather magnetic tape which is electronically inscribed. This has a number of advantages over the still camera. First, videocams can operate in at least one candlelight brightness without accessory lights or flashes and, second, unlike standard film, you can replay your shoot instantly. This replay occurs in miniature form in the viewfinder. If you don't like what you see you can re-shoot and tape over the previous shot.

A further interesting feature about the video camera is that it can record sound as well as image. The microphones are usually 'shot gun' directional, that is, they have a narrow but extremely sensitive focus directly in front of the lens. They are not unidirectional and this means that sound at the sides of the camera is missed or is heard as minor and distant.

For the unobtrusive researcher, a truly interesting feature of many video cameras is the adjustable viewfinder. The operator can use the viewfinder either with the lens pointing directly in front or, alternatively, the camera can sit sideways with the lens pointing to the left or the right of the operator. This design feature means that you don't have to look as if you are filming. You can sit side on from the scene and hold a conversation, all the while taping the scene to the side of your seated position. However, there are several problems of cost. In 1992 in Australia, video cameras cost between $1300–$2000 while a simple fully automatic still camera usually costs around $200 or less. In addition to the initial outlay is the video playback equipment and batteries. In the case of batteries, these tend to last from 45–90 minutes each before recharging. This means that it is advisable to carry at least two or three with you in the field. At 1992 prices of around $80 each, these do not come cheaply. If you are not supplied this equipment through an institution (such as a university or company) or you cannot buy the equipment through the proceeds of a research grant, perhaps you might think of hiring rather than buying equipment.

Whatever you decide, you will need to practise for a couple of hours to get used to the equipment but its automation and simplicity ensures that you should not find it difficult to use in observational work. With all equipment, be careful in extremes of weather. Protect all cameras from water with waterproofing plastic. In areas of high humidity (for example, North Queensland or parts of the Northern Territory) the video tape may jam. In that case the cassette housing will simply need airing for a few minutes. If you are a newcomer to this equipment do not leave home without the manufacturer's trouble-shooting guide. Without this, simple problems can become time expensive for you.

Advantages and disadvantages

Gary Albrecht (1985) wrote an extremely valuable article on the uses of video camera work in social science research, and this is a short and useful preliminary reading for anyone contemplating this piece of hardware. Albrecht (1985) and Werner and Schoepfle (1987) discuss the strengths and problems of video camera work. The value of using videos is that tapes are re-usable, they are easily duplicated, can be previewed immediately, are easily transportable and they have sound capability. They are useful behind one-way mirrors, filming at a distance and recording crowd behaviour. Speeding up and slowing down the film allows details to be seen in fresh ways. Behaviour analysis can proceed from seconds through to hours. Repeat viewing facilitates more detailed analysis of complex events and sequences. Finally, multiple cameras used simultaneously allow divergent perspectives to be recorded of the one behaviour, event or process.

The disadvantages include bias due to poor sampling; reactivity, if poor handling of equipment attracts attention; distortion of images if lenses are improperly used (for example, getting too close to a subject with wide angle). The safeguard that Werner and Schoepfle (1987, p. 283) recommend is to view all photographs or film with minimum editing. Unfortunately, this will not necessarily overcome problems of distortion and bias. Furthermore, if a large number of photographs or considerable film footage has been taken, this safeguard becomes rather unwieldy. For short photographic assignments, a detailed description, verbal or written, of the sampling, and field and equipment procedures, are the best methods for

Advantages of photography (still and video)

- Easily transportable
- Permits extremely detailed analysis
- Easy to re-analyse
- Permits inter-rater checks for reliability
- Easily learnt

evaluating the problems, however dull and literary these may appear.

Disadvantages of photography (still and video)

- Open to the usual problems of observer bias
- Can attract unwanted attention
- Lens can distort shape/size
- Can be expensive
- Some operating difficulties exist in extreme climates

I have deliberately not discussed the issue of film developing for still camera work or editing for video camera work for two main reasons. First, film developing and editing require highly complex skills which are often beyond the newcomer. For those who wish to make photography their main methodological interest in life, there are many good books, guides and short courses available. That task is beyond the scope of this book. Second, developing and editing equipment is specialised, expensive and at times dangerous. Warning you about methodological pitfalls is as far as this book goes. Providing guidance about which chemicals you should *not* put together is someone else's calling in life. I suggest you consult the relevant reference material and bank account for this equipment.

Ethics

The ethical problems one might consider in the use of photography as a method are the following. First, be aware that you might inadvertently record crime. Albrecht mentions drug deals, car theft rackets and illegal business transactions that he and others have recorded by accident. This is not to mention the recording of embarrassing information such as sexual liaisons between people thought previously to be faithful to one partner or sexual orientation. Confidentiality and the protection of privacy are the main considerations here. However, what if your film records a rape? If, by enlarging the photograph or slowing the film the rapist can be identified, then the relevant authorities should be notified. Clearly, these issues need careful thought as Albrecht says, 'on a case by case basis with the needs of the litigant' in mind. (1985, p. 337).

Unobtrusive recording in public places is not ethically objectionable because simple observation in public settings is not. Since film is materially different from written notes, however, Albrecht advises that either subjects give consent (if there are only three or four) and/or that confidentiality is strictly observed and that records be destroyed after use. Of course, the people photographed must not be 'appreciably affected by the study' either and that means that the film should not fall into the hands of journalists and investigators. In this connection, Asch (1988) warns that filming in foreign countries can attract the attention of some governments. Some officials may insist on seeing all the exposed or recorded film and this may be extended to a request for full copies of everything. When governments make these kinds of requests one may reasonably reflect on the harmful potential such a requirement might have on those appearing on your prints or film. Choose your country carefully.

What do I photograph?

Remember that cameras are hardware, not method. The method is photography or photographic observation and the camera is the recording device. Therefore you should first study the steps to simple observation in chapter seven. The first step requires you to define your interest. If you are interested in expressive movement or objects, select your camera, lens and film accordingly. With the photography of objects, shape and colour can be important so remember to photograph in good light and perhaps take several shots to show shape. Close up is good for detail but unless surrounding details are also photographed the impression of size and context (important physical and sociological considerations) will be lost.

Collier and Collier (1986, p. 41) offer some useful guidelines for ethnographic photography. Some of these are summarised on page 150 but for an extended discussion consult the original work.

A similar system using still photography was used by Dona Schwartz (1989) in her study of change in a rural farm community in the USA and I recommend a look at this article for her methodology. Schwartz used her photographs as the basis for later interviews.

Whatever your interest, whether this be expressive movement or the more ambitious ethnographic portrayal of some group's social life, the issues of sampling and recording, planning and analysis as

Photographic perspectives include:

- location—boundaries, landmarks, geographical features, signs etc.
- appearance—visual character such as buildings, streets, hills and flats, winding and straight etc.
- organisation—where businesses, churches, public places, residences are located. 'Make wide shots to define relative location, close-ups that show details' (p. 41)
- functions—range of social activities, services, sub-cultures, occupational groups
- people—ages, ethnic groups, range of people seen, visitors etc.
- transport—roads, paths, trains, trams, car parks, congestion, who uses what
- residences—character and condition, age, class, ethnic groupings in these
- daily cycles—happening where and when, and photos of same area at different times
- history—reflections of the past, buildings, population and their impact as well as incongruencies
- change—what is changing and what is not, closings and openings, old people, newcomers

(Adapted from Collier and Collier, 1986, p. 41.)

discussed in chapter seven apply. Work through those steps carefully.

Computer modelling

If you are like me, you use your computer for two purposes, as a paper weight and as a video typewriter. The good news about computer modelling is that, in abstract terms, it is quite easy to explain. The bad news is that, in practical terms, it is extremely difficult to do. I will therefore begin with the easier task.

Modelling is the trial run of an idea without putting it into actual practice. Rather than attempting to evaluate the problems of a design or course of action by actually creating or embarking on them, a computer is used to simulate or model the event. Scale models of boats or cars, for example, are used to assess the practical problems of full-size vehicles. Mathematical models are used to test physical ideas in abstract ways. In computing, you place all the known or expected inputs or variables into the model and then allow the computer to calculate the outcome or consequences.

For example, some simple arithmetic can act as a model for what happens when you adjust the price of a certain product, say, water filters for household taps. If ten of these filters are sold every day when their price is one dollar, how many will sell when you raise the price to five dollars. Perhaps in that scenario few filters will be bought. However, if the media reported that asbestos fibres had been found in domestic drinking water, filter sales may sky-rocket. If the media reported that domestic water contained weight-reducing fibres which were taken out by filters, you may not be able to give water filters away.

Calculating simple models that contain relatively unchangeable inputs, such as determining the effect of a constantly rising temperature at a certain rate, over a certain time period, upon a known physical substance, is called a 'deterministic' model. More complex models which are design to rehearse multiple 'what if' scenarios and variables are known as 'stochastic' models.

Models are widely used to test and develop theories of one kind or another. For example, Meadows and Robinson (1985) describe a wide variety of agencies which use modelling: the World Bank, universities, business, forecasting groups and health and welfare groups are among those that use simulation experiments to assess a diverse range of problems. These can be complex models of economic growth or rates of development and inflation. They can also include agricultural development and commodity market analysis and prediction, weather forecasting, and the assessment of diverse patterns of population growth, famine, drought or desertification on resource issues.

Business research has found modelling particularly useful for budgeting, cash flow analysis, personnel and profitability planning, marketing, manufacturing and inventory analysis. The financial and marketing side of computer modelling has had far-reaching impact on the operations of companies. A dramatic example of this is cited by Behan and Holmes (1986, 404–6) in their discussion of the role of computer simulation in the business affairs of the Nestlé Company.

Aside from business, computer modelling has also been used, albeit much less so, in a variety of humanities and social science disciplines. See, for example, the historical study employing modelling to analyse Caesar's Gallic campaign (Ricketts, 1978) or the archaeological study of house construction in the Bronze Age (Walsh, 1982). A careful browse through the library catalogues under 'computers' or 'computer modelling and simulation' will turn up an abundance of examples of this type. I will provide one further but more detailed example of computer modelling in the service of the social biologist Richard Dawkins (1986).

Dawkins attempted to support his particular theory of evolution by showing how cumulative rather than single-step selection is the key to understanding how evolution works. Most people think that evolution is about blind chance (i.e. single selection) but in fact, as Dawkins argues, evolution is possible and credible only because of the existence of cumulative selection. Dawkins uses the example of a monkey attempting to reproduce, on a typewriter, the Shakespearian line '*Methinks it is like a weasel*'. That line has twenty-eight characters (including spaces). For each 'try' our hapless monkey is permitted twenty-eight random bashes at the typewriter. This is single-step selection. At this rate, the monkey may arrive at the Shakespearian line in 'about a million, million, million, million, million years. This is more than a million, million, million times as long as the universe has so far existed' (p. 49).

On the other hand, cumulative selection compares its first 'try', its first randomly chosen twenty-eight characters, with the target ideal. The next selection, still random but now purposive, is based on the information gained from the first. In this fashion, the Shakespearian line takes eleven seconds to appear on the computer screen. Dawkins uses his personal computer to develop these exercises. He goes on with this experiment by developing, not letters and sentences, but rather graphics or images.

Following these simple principles Dawkins developed a program which was designed to simulate organic or genetic development using a simple recursive technique which produced line drawings of tree-like characters. These programs were linked with others that shaped or changed this development. This 'gene' program simulated reproduction and mutation. With the development and reproduction programs simulating genetic growth and development, there remained only the problem of simulating natural (cumulative) selection which would act as the 'target ideal' influence over that growth and development. Here Dawkins' own eye would do the selecting. As each screening of the programs offered a choice of graphics and images which resembled insects or animals, Dawkins would select the images that most interested him in an arbitrary way. Like nature and natural selection itself each choice was 'opportunistic, capricious and short term'. The 'target ideal' here regularly shifted as indeed does natural selection itself. The result was pages of graphics which resembled scorpions, bats, foxes, frogs, spiders and even butterflies, people's faces and lamps! And all these from an initial program which produced little stick-like tree graphs. Programmed to mutate, and selected for further mutation on some arbitrary criteria of aesthetics, Dawkins' simulation provided powerful and persuasive evidence for his theory of natural selection and genetic evolution.

Advantages and disadvantages

The advantages of computer modelling as an unobtrusive method are first, that it is a powerful theory building tool. It allows researchers to test the strength and meaning of certain links, associations and interactions. In quantitative, hypothetico–deductive studies, computer modelling can be rigorous, controlling for variables in the same way as an experimental design. Unlike the camera, computer programming can be an inexpensive pastime, although initial costs can be quite high. Finally, as Whicker and Sigelman (1991, p. 131) remark, computer modelling can 'allow worst case scenarios (for example, surviving a nuclear attack) to be explored without really suffering the worst case conditions'.

Advantages of computer modelling:

- It allows theory building and testing.
- It is rigorous for hypothetico–deductive designs.
- It is inexpensive after initial costs.
- It permits worst case scenarios to be tested without researcher experiencing the actual worst case conditions.

The disadvantages of modelling are that simulation is wholly dependent on the accuracy and appropriateness of the input variables and assumptions made by the researcher/programmer. Second, modelling does not handle ambiguity well. All general processes or experiences must be programmed into a model in logical and quantifiable ways if the program is not to become over-burdened with 'bugs', that is, command propositions which either do not work or are unable to conclude their operations satisfactorily. Commands for the program must be highly specific and logical. Finally, program skills can be difficult and time consuming to learn. The jargon and logic of programming language can be impenetrable to the novice/newcomer.

Disadvantages of computer modelling:

- Dependent upon the assumptions and information of the programmer
- Processes only highly specific and logical commands and has poor tolerance of ambiguity
- Skills required are time consuming and difficult to learn

Whicker and Sigelman (1988) provide a good introduction to this area (see especially chapter eight) and they also provide a useful annotated bibliography of practical guides to simulating and modelling for those who already understand the basics of programming language and technique.

Programming for the ignorant

If programming for modelling and simulation looks exciting as an avenue of unobtrusive research but you don't know where to start, here are a few suggestions. First, once you have a computer at your disposal (most types of contemporary personal computer are fine), you will need to choose a 'language' with which to program your computer. The most common software languages are FORTRAN, which is mainly used by physical scientists (for example, in weather forecasting) and academics using large university 'mainframe' computers and 'C' language which is probably the most common programming language to use. Well-known software packages in 'C' language are MICROSOFT WORD and MICROSOFT WRITE. Many of us would have come across these during our word processing tasks. Finally, there are languages such as SIMSCRIPT. This is a higher level language that is especially adapted for modelling. It needs fewer commands than 'C' language for example and this, sometimes reduces the input ten-fold. Consequently it is much quicker and easier to learn. SIMSCRIPT and similar modelling languages also have the option of exiting into 'C' language if the need arises.

To be really proficient in programming (and programming to the point of programming computer models in 'C' language) can take many years. If one concentrates on SIMSCRIPT for example, learning can still take quite a few months. A more gentle path to programming language would be to learn MICROSOFT Q BASIC. This is usually provided free with all DOS.5 computers (IBM and IBM compatible computers). MacIntosh equipment has an equivalent language in its software offerings. Unlike cameras, and contrary to the view of many computer zealots, working with computers is not a user-friendly pastime—especially for the bedrock ignorant.

Nevertheless, somewhere in every city, in some tertiary institution or company, evening courses are held to train the bold and brave in these different computer languages (particularly MICROSOFT Q BASIC language). The costs can range from a couple of hundred dollars for a semester to a thousand dollars for privately run crash courses over several evenings.

For those more timid among us, you will be pleased to know that computer models can be bought over the counter at your local computer supplier. The TREASURER GAME (Perkins, 1992: $69 r.r.) simulates the Australian economy. You are not able to change the underlying model but you can change the inputs (for example, the tax rate, sales tax, imports etc.). This is used in some university courses and rumour has it that the Federal Treasurer owns one of these too. SIM CITY (1989–91) is a model of a city. It is used for analysing the input resources supplying schools, for example utilities such as gas, water and electricity, health and welfare institutions and so on. Increasing the resource input to one area affects the welfare of another area of the city.

Anyone can play with these models and observe the outcomes. The problem here, of course, is that you are confined to the assumptions of the creator of the model. Some models do allow alteration of the model's assumptions by the introduction of new commands. However, usually the creators do not guarantee that the program will work or solve the problem to which it is assigned.

Ethics and modelling

Because computer models simplify variables so that these can be manipulated more easily there is always a danger that models can *mislead.* This can occur in two ways. First, deliberately leaving out variables or influences. One can be less than honest with oneself

Requirements for computer programming include:

- at least, one personal computer
- a programming language, for example Q BASIC, PASCAL, 'C' or SIMSCRIPT
- attendance at a TAFE/university/private course of instruction on programming *or*
- the purchase of a commercially produced model of your choice to help create better or clearer models in your head

about the possible importance of certain influences thus leading others to interpret the model poorly due to a lack of information and understanding of how the basic assumptions of the model have been derived. Second, irrespective of whether or not one is dishonest or simply naive about certain influences, not accurately disclosing those assumptions when revealing the model to others can mislead them into false confidence in the model. If social policies or actions are based on this model, and the model proves deficient in important ways, harm may result for those people affected by those policies or actions.

On the other hand, one may also ask whether it is ethically responsible to forecast, through modelling, other people's futures and then attempt to manipulate the outcome. People do this regularly in stock exchange forecasts but how ethically sound is this for economic theory or social welfare policy? Computer modelling put to social uses raises all the usual ethical problems of social models and policies but one more. Because people associate the computer with 'science' and 'objectivity', social models or policies developed inside them can obscure the fact that, in the first instance, the computer is programmed with human assumptions. Those assumptions can be easily forgotten or conveniently laundered or overlooked in any rhetoric about scientific modelling and computer research in human affairs. Researchers should be careful not to exploit those popular images by always providing clear qualifications and limitations about the models they develop for others.

Conclusion

Computers and cameras are pieces of equipment that help you to see and think about the world in new ways. They are not, however, any different from other technologies such as the microwave oven.

In that sense, the quality of what you get out of this technology is highly dependent upon what you put in. Cameras and computers do not create data any more than ovens create dinner. But these technologies make you conscious of your own ways of seeing and thinking by linking 'what you see' with 'who you are'. For example, the 'bugs' in a computer program often occur because *your* commands or *your* logic has been less than precise. If the program arrives at conclusions or scenarios that are unlikely or incredible, you may have stumbled upon a wonderful discovery or have run a program with hopelessly unlikely assumptions. Both outcomes are based on your view, your logic, your assumptions.

Similarly, photos are a product of *your* own categories of experience and the types of events and people which attract *your* eye. This quality of making you self-conscious about how you think and see the world really is the great strength of hardware and sets it apart from other techniques available to the unobtrusive researcher. By controlling for, and being reflective about, the limits of your perceptions, social research has the possibility of transcending them, of going beyond them and surprising you. Cameras and computers in that sense are aids to imagination and 'when we are prevented from making a journey in reality, the imagination is not a bad substitute' (Dawkins, 1986, p. 74).

Recommended reading

Albrecht, G.L. (1985), 'Videotape safaris: entering the field with a camera', *Qualitative Sociology*, vol. 8, no. 4, pp. 325–44

Ayers, R., Mollison, M., Stocks, I. and Tumeth, J. (1992), *Guide to Video Production*, Allen & Unwin, Sydney.

Behan, K. and Holmes, D. (1986), *Understanding Information Technology*, Prentice-Hall, Sydney

Collier, J. and Collier, M. (1986), *Visual Anthropology: Photography as a Research Method*, University of New Mexico Press, Albuquerque (see especially chapter 19)

Hawkins, A. and Avon, D. (1979), *Photography: Guide to Technique*, Blandford Press, London. (A good book for beginners, though in this category there are many to choose from. Some published by National Geographic and Kodak Eastman are also worth browsing.)

Neelamkavil, F. (1987), *Computer Simulation and Modelling*, John Wiley, Chichester

Rollwagen, J.R. (1988), *Anthropological Filmmaking*, Harwood Academic Publications, London.

Schwatz, D. (1989), 'Visual ethnography: using photography in qualitative research', *Qualitative Sociology*, vol. 12, no. 2, pp. 119–54

Whicker, M.L. and Sigelman, L. (1991), *Computer Simulation Applications*, Sage, Newbury Park

Ziller, R.C. (1990), *Photographing the Self: Methods for Observing Personal Orientations*, Sage, Newbury Park (see especially the first two chapters)

A parting note

Anthropologists, psychologists and sociologists have been keen to talk, particularly the question and answer kind of talk. It is true that observation has played a role in social research but this has often been either of animals (for example, rats, pigeons, monkeys) or of people from 'foreign' places (as part of the Australian colonial past). Rarely has both talk and image come together as in real life, and indeed as in modern film, for social science research. When research was not decontextualised (i.e. occurring away from the actual events being discussed) it was commonly disembodied (i.e. the results were presented in abstract numerical tracts that denied the richness of the original experience).

The drawcard of qualitative interview methods has been precisely in their attempt to overcome the disembodied and decontextualised ways of survey and experimental work in social research since World War II. But, ironically, that attempt has served to fetishise and concentrate undue attention on the spoken word, much like the quantitative obsession with numbers. I say 'undue' because people other than writers, academics, professional and clerical groups place a lot less emphasis on the spoken word. And herein lies the key reason why talkative, interrogative methods of social research are not enough. The talk-only approach to studying any aspect of human culture is strangely academic and ethnocentric in two ways.

First, in many cultures and times, the spoken word has proved inadequate for its user's purposes and such problems have served as springboards for other expressions—art, music, dance, theatre, ritual, touch, looks, gestures, onomatopoeia, costume, scribble, memorabilia and many other examples. When words have been seen as important, context has been critical to their understanding by others. Unobtrusive methods then, while drawing our sentimental attention away from the spoken word and towards the object, the image, the behaviour or the memoir, free up the word-bound nature of our social science record of human experience. This prompts an exciting comparison, I think, between the seen and the said and also between the material and the symbolic.

Second, the talk-only approach has become increasingly technical and inaccessible to all but a few academics and professionals. There is a growing and somewhat unchecked view that all interviews or survey data should be subjected to complex computation or statistical process before being recognised as 'real' research. On the other hand, unobtrusive methods can be undertaken by anyone. In most cases, most people can learn to look and record systematically and provide an analysis based on reading and reason. In this respect, many of the methods and sources are wonderfully accessible. People look at each other and their cultural products and activities and they attempt to explain these relationships. Other people may agree or disagree with those findings, but in the process they too contribute to a discourse—a big conversation in society. That discourse does not rest solely on confessions of belief, attitude and motive but on the deep, broad, overlooked, taken-for-granted world of the visual. When the visual is explored, either in written records or current behaviour, it can provide, by itself or in combination with survey and interview results, novel questions and broader perspectives. Research which is less academic or less technical in appearance is not necessarily any less useful. On the contrary, just as qualitative in-depth interviews showed up the sterility and uni-dimensionality of social statistics, so, too, unobtrusive methods expose the folly of relying solely on what others say to understand human culture in any of its diverse settings.

Finally, unobtrusive methods are important to contemporary social researchers because these methods trace the interrelationship of ideas to their material and behavioural expressions and influences. Objects, settings, behaviours and ideas are discursive, they reflect and feed off each other. Ideas are products, not simply of minds, but minds-in-context. In concert with other methods, unobtrusive methods can provide relevant social sources and techniques to recover that context. Beliefs and attitudes can be matched with actual behaviour, lifestyle or habitat and that is a further strength of this methodology. Current ideas may also be checked, traced and contextualised by the past. And when a historical context is sought, and particularly in the absence of anyone to talk to, unobtrusive examination of literary, visual or material sources can be the only passage to the past, the only clues, however modest, to that context. Although when used alone unobtrusive methods have as many problems as any other method used alone, they are still nevertheless able to make a unique contribution. Unobtrusive methods are able to make the silent past accessible to researchers of the present. The unobtrusive researcher is able to enter places, see people and events, and listen to voices that pollsters and interviewers are unable to hear, see or visit with their methods.

In the 1960s when Webb and his colleagues' book was released, it was seen as a novelty item, an interesting sideshow in the circus of mainstream methodology. In the 1990s I hope you will have seen in these pages the fact that unobtrusive methods are not simply novel. They are not merely additives or optional extras to surveys or interviews but rather a collection of rich techniques and sources which help portray the broader social picture of human activity and experience. These are important elements to capture and appreciate.

These methods may be basic and simple in their representation as introduced here but they can start you in a very promising direction. And whatever might be their problems, they can always serve to remind you that if you really want to conduct a serious study of humanity, you must use everything you've got, to look at everything we've got.

Bibliography

Akeret, R.V. (1973), *Photoanalysis: How to Interpret the Hidden Psychological Meaning of Personal and Public Photographs*, (ed. T. Humber), Wyden, New York

Albrecht, G.L. (1985), 'Videotape safaris: entering the field with a camera', *Qualitative Sociology*, vol. 8, no. 4, pp. 325–44

Allen, R.C. (ed.) (1987), *Channels of Discourse*, Routledge, London

Altman, R. (1984), 'Television/Sound', paper delivered at the 24th Annual Meeting of the Society for Cinema Studies, Madison, 24 March

Anonymous (1973), *Go Ask Alice*, Corgi, London

Asch, T. (1988), 'Collaboration in ethnographic filmmaking: a personal view', in J.R. Rollwagen (ed.) *Anthropological Filmmaking*, Harwood Academic, London, pp. 1–29

Ayers, R., Mollison, M., Stocks, I. and Tumeth, J. (192), *Guide to Video Production*, Allen & Unwin, Sydney

Babbie, E. (1989), *The Practice of Social Research*, 5th edn, Wadsworth, Belmont, CA

Barthes, R. (1985), *The Fashion System*, Jonathan Cape, London

Bath, J.E. (1981), 'The raw and the cooked the material culture of a modern supermarket', in R.A. Gould and M.B. Schiffer (eds), *Modern Material Culture: The Archaeology of Us*, Academic Press, New York, pp. 183–95

Bayley, S. (1986), *Sex, Drink and Fast Cars*, Faber & Faber, London

Beaumont, J. (1988), *Gull Force: Survival and Leadership in Captivity 1941–1945*, Allen & Unwin, Sydney

Behan, K. and Holmes, D. (1986), *Understanding Information Technology*, Prentice-Hall, Sydney

Bell, C. and Encell, S. (1978), *Inside the Whale: Ten Personal Accounts of Social Research*, Pergamon, Rushcutters Bay, NSW

Barnard, S. (1988), *On the Radio: Music Radio in Britain*, Open University Press, London

Berry, J.W. (1979), 'Unobtrusive measures in cross-cultural research', in L. Sechrest (ed.) *Unobtrusive Measures Today*, Jossey-Bass, San Francisco

Blake, C.F. (1981), 'Graffiti and racial insults: the archaeology of ethnic relations in Hawaii', in R.A. Gould and M.B. Schiffer (eds) *Modern Material Culture: The Archaeology of Us*, Academic Press, New York, pp. 87–99

Bochner, S. (1979), 'Designing unobtrusive field experiments in social psychology' in L. Sechrest (ed.) *Unobtrusive Measures Today*, Jossey-Bass, San Francisco

Bourdieu, P. (1990), *Photography: A Middle-Brow Art*, Polity Press, Cambridge

Brandt, R.M. (1972), *Studying Behaviour in Natural Settings*, Holt, Rinehart and Winston, New York

Brooks, P. C. (1969, *Research in Archives*, University of Chicago, Chicago

Brown, D. (1970), *Bury My Heart At Wounded Knee: An Indian History of the American West*, Holt, Rinehart and Winston, New York

Bruner, E.M. and Kelso, J.P. (1980), 'Gender differences in graffiti: a semiotic perspective', *Women's Studies International Quarterly*, vol. 3, pp. 239–52

Bulmer, M. (ed.) (1984), *Sociological Research Methods: An introduction*, Macmillan, London

Burnstein, S., Goodhew, V., Reed, B. and Tranter G. (eds) (1992) *Directory of Archives in Australia*, Archivists Society of Australia, Canberra

Burton, G. (1990), *More Than Meets The Eye*, Edward Arnold, London

Carney, T.F. (1972), *Content Analysis*, Batsford, London

Carpenter, J.P. (1977), *The Screwdriver (Does It or Doesn't It?)*, Tucson, Arizona State Museum Library, University of Arizona

Clegg, S. (1990), *Organisational Studies in the Postmodern World*, Sage, London

Cleghorn, P. L. (1981), 'The community store', in R.A. Gould and M.D. Schiffer (eds), *Modern Material Culture: The Archaeology of Us*, Academic Press, New York, pp. 197–212

Codlin, E. (ed.) (1990), *Directory of Information Sources in the U.K.*, Aslib, London

Collier, J. and Collier, M. (1986), *Visual Anthropology: Photography as a Research Method*, University of New Mexico Press, Alburqueque

Colquhoun, D. (1990), 'Images of healthism in health-based physical education', in D. Kirk and R. Tinning (eds), *Physical Education, Curriculum and Culture: Critical Issues in the Contemporary Crisis*, The Falmer Press, London

Connolly, J. and Keutner, T. (eds) (1988), *Hermeneutics Versus Science*, University of Notre Dame Press, Notre Dame

Cooke, K. (1992), 'Keep yourself nice' (Letters), *The Age*, 25 January, p. 16 Extra

Covey, H.C. (1991), *Images of Older People in Western Art and Society*, Praeger, New York

Curry, T.J. and Clarke, A.C. (1983), *Introducing Visual Sociology*, Kendall/Hunt, Dubuque, Iowa

Dale, A., Arber, S. and Procter, M. (1988), *Doing Secondary Analysis*, Allen & Unwin, London

Daly, J. (1993), 'Team research in clinical settings: strategies for qualitative researchers', in D. Colquhoun and A. Kellehear (eds), *Health Research in Practice*, Chapman and Hall, London, pp. 24–36

Daniel, A. (1983), *Power, Privilege and Prestige: Occupations in Australia*, Longman Cheshire, Melbourne

Davis, F. (1989), 'Of maids uniforms and blue jeans: the drama of status ambivalences in clothing and fashion', *Qualitative Sociology*, vol. 12, no. 4, pp. 337–55

Dawkins, R. (1986), *The Blind Watchmaker*, Longman Scientific and Technical, Harlow

de Vaus, D. (1990), *Surveys in Social Research*, Unwin Hyman, London

Deeley, J. (1990), *Basics of Semiotics*, Indiana University Press, Bloomington

Deetz, J. (1977), *In Small Things Forgotten*, Anchor Press, New York

Denzin, N. (1970), *The Research Act*, Aldine, Chicago

—— (1991), *Images of Postmodern Society*, Sage, London

Derrida, J. (1981), *Positions*, Athlone, London

Dethlefsen, E.S. (1981), 'The cemetery and culture change: archaeological focus and ethnographic perspective', in R.A. Gould and M.B. Schiffer (eds), *Modern Material Culture: The Archaeology of Us*, Academic Press, New York, pp. 137–59

Dissertation Abstracts International, University Microfilms, Ann Arbor, Michigan

Douglas, J.D. (1967), *The Social Meanings of Suicide*, Princeton University Press, Princeton, New Jersey

Dowdall, G.W. and Golden, J. (1989), 'Photographs as data: an analysis of images from a mental hospital', *Qualitative Sociology*, vol. 12, no. 2, pp. 183–213

Dunkin, M. (1992), 'Some dynamics of authorship', *The Australian Universities Review*, vol. 35, no. 1, pp. 43–8

Durkheim, E. (1951), *Suicide: A Study in Sociology*, The Free Press, New York

Dyer, G. (1982), *Advertising as Communication*, Methuen, London

Eagleton, T. (1983), *Literary Theory: An Introduction*, Blackwell, Oxford

Eibl-Eibesfeldt, I. (1975), *Ethology: The Biology of Behaviour*, Holt, Rinehart and Winston, New York

Eichler, M. (1988), *Nonsexist Research Methods: A Practical Guide*, Allen & Unwin, Winchester, Massachussets

Elkin, A.P. (1945), *The Australian Aborigines: How to Understand Them*, Angus & Robertson, Sydney

Ellis, J. (1982), *Visible Fictions: Cinema, Television, Video*, Routledge & Kegan Paul, London

Fields, E.E. (1988), 'Qualitative content analysis of television news: systematic techniques', *Qualitative Sociology*, vol. 11, no. 3, pp. 183–93

Figlio, K. (1982), 'How does illness mediate social relations: workman's compensation and medico-legal practices 1899–1940', in P. Wright and A. Treacher (eds), *The Problem of Medical Knowledge: Exercising the Social Construction of Medicine*, Edinburgh University Press, Edinburgh

Fiske, J. (1987), *Television Culture: Popular Pleasures and Politics*, Methuen, London

Fook, J. (1991), 'Is casework dead?: a study of the current curriculum in Australia', *Australian Social Work*, vol. 44, no. 1, pp. 19–28

—— (1993), *Radical Casework*, Allen & Unwin, Sydney

Ford, L. (1971), 'Geographic factors in the origin, evolution and diffusion of rock and roll music', *Journal of Geography*, vol. 70, no. 8, pp. 455–64

Foucault, M. (1972), *The Archaeology of Knowledge*, Tavistock, London

Frank, A. (1954), *The Diary of Anne Frank*, Pan Books, London

Frith, S. (1986), 'Art vs. technology: the strange case of popular music', *Media, Culture and Society*, vol. 8, pp. 263–79

Game, A. (1991), *Undoing the Social: Toward a Deconstructive Sociology*, Open University Press, London

Geertz, C. (1973), *The Interpretation of Cultures*, Basic Books, New York

Gilbert, L. (1980), *A Grave Look at History*, John Ferguson, Sydney

Glassner, G. and Moreno, J.D. (1989), *The Qualitative–Quantitative Distinction in the Social Sciences*, Kluwer Academic Publishers, Dordrecht

Goffman, E. (1979), *Gender Advertisements*, Harper & Row, New York

Goode, E. (1982), 'Multiple drug use among marijuana smokers' in G. Rose (ed.), *Deciphering Sociological Research*, Macmillan, London, pp. 192–206

Gould, R.A. and Schiffer, M.B. (eds) (1981), *Modern Material Culture: The Archaeology of Us*, Academic Press, New York

Gregory, C.A. and Altman, J.C. (1989), *Observing the Economy*, Routledge, London

Gross, D. (1984), 'Time allocation: a tool for the study of cultural behaviour', *Annual Review of Anthropology*, vol. 13, pp. 519–58

Haigh, G. (1992), 'History up in smoke', *The Age*, 15 February, p. 7 Extra

Ham, C., Nettl B. and Byrnside, R. (1975), *Contemporary Music and Music Cultures*, Prentice-Hall, Englewood Cliffs, New Jersey

Harper, R.G. and Wiens, A.N. (1979), 'Non-verbal behaviours in unobtrusive measures', in L. Sechrest (ed.), *Unobtrusive Measures Today*, Jossey-Bass, San Francisco

Hawkins, A. and Avon, D. (1979), *Photography: Guide to Technique*, Blandford Press, London

Heaven, P.C.L. (1991), 'Voting intention and the two value model: a further investigation', *Australian Journal of Psychology*, vol. 43, no. 2, pp. 75–7

Hodder, I. (1989), 'This is not an article about material culture as text', *Journal of Anthropological Archaeology*, vol. 8, pp. 250–69

Holsti, O.R. (1969), *Content Analysis for the Social Sciences and Humanities*, Addison-Wesley, Reading, Massachusetts

Hood, M. (1971), *The Ethnomusicologist*, McGraw-Hill, New York

Humphrey, L. (1975), *Tearoom Trade: Impersonal Sex in Public Places*, Aldine de Gruter, New York

Huntingford, F. (1985), *The Study of Animal Behaviour*, Chapman & Hall, London

Insinger, M. (1991), 'The impact of a near-death experience on family relationships', *Journal of Near-Death Studies*, vol. 9, no. 3, pp. 141–82

Irwin, D. (1981), 'Sentiment and antiquity: European tombs 1750–1830', in J. Whaley (ed.) *Mirrors of Mortality*, Europa, London, pp. 131–53

Kellehear, A. (1989), 'Ethics and social research', in J. Perry, *Doing Fieldwork: Eight Personal Accounts of Social Research*, Deakin University Press, Geelong, pp. 61–72

Kellehear, A. (1990), *Dying of Cancer: The Final Year of Life*, Harwood Academic Publishers, London

—— (1990), 'A critique of national health policies for country people', in T. Cullen, P. Dunn and G. Lawrence (eds), *Rural Health and Welfare in Australia*, Charles Sturt University–Riverina, Wagga, pp. 28–37

Kern, A. (1970), 'The sociology of knowledge in the study of literature', in M.C. Albrecht, J.H. Barrett and M. Griff (eds), *The Sociology of Art and Literature*, Praeger, New York, pp. 553–61

Kirk, J. and Miller, M.L. (1986), *Reliability and Validity in Qualitative Research*, Sage, Newbury Park

Klofas, J.M. and Cutshall, C.R. (1985), 'The social archaeology of a juvenile facility', *Qualitative Sociology*, vol. 8, no. 4, pp. 368–87

Knop, E. (1969), 'Suggestions for interpreting articles', in M. Abrahamson (ed.), *Introductory Readings on Sociological Concepts, Methods and Data*, Van Nostrand, New York, pp. 67–71

Koutroulis, G. (1990), 'The orifice re-visited: woman in gynaecological texts' *Community Health Studies*, vol. 14, no. 1, pp. 73–84

Kristeva, J. (1987), *Tales of Love*, Columbia University Press, New York

Lagace, R.O. (1974), *Nature and Use of HRAF Files: A Research and Teaching Guide*, Human Relations Area Files, New Haven, Connecticut

Lefebvre, H. (1971), *Everyday Life in the Modern World*, Allen Lane London

Leunig, M. 'The wall', *The Age*, 4 January, p. 8 Extra

Lofland, J. (1971), *Analyzing Social Settings: A Guide to Qualitative Observation and Analysis*, Wadsworth, Belmont, California

Lomas, A. (1968), *Folk Song Style and Culture*, American Assoc. for the Advancement of Science, Washington

Lupton, D. (1992), 'Discourse analysis: a new methodology for understanding ideologies of health and illness', *Australian Journal of Public Health*, vol. 16, no. 2, pp. 145–50

Luscher, M. (1971), *The Luscher Color Test*, Pocket Books, New York

Lyn, J. and Jay, A. (1984), *The Complete Yes Minister: Diaries of a Cabinet Minister*, British Broadcasting Corporation, London

Lyotard, J.F. (1984), *The Postmodern Condition: A Report on Knowledge*, Manchester University Press, Manchester

McQueen, D.R. (1973), *Understanding Sociology Through Research*, Addison-Wesley, Reading, Massachusetts

Maher, C. and Burke, T. (1991), *Informed Decision Making*, Longman Cheshire, Melbourne

Mannheim, K. (1936), *Ideology and Utopia*, Routledge & Kegan Paul, London

Manning, P. K. (1987), *Semiotics and Fieldwork*, Sage, Beverly Hills

Martin, P. and Bateson, P. (1986), *Measuring Behaviour: An Introductory Guide*, Cambridge University Press, Cambridge

Mayer, G. and Burton, L. (1991), *Media Studies*, Jacaranda Press, Milton

Meadows, D.H. and Robinson, J.M. (1985), *The Electronic Oracle: Computer Models and Social Decisions*, Chichester, New York

Metz, C. (1974), *Film Language: A Semiotics of the Cinema*, Oxford University Press, New York

Middleton, R. (1981), *Form and Meaning: Reading Popular Music*, Open University Press, Milton Keynes

Middleton, R. (1990), *Studying Popular Music*, Open University Press, London

Miles, M.B. and Huberman, A.M. (1984), *Qualitative Data Analysis*, Sage, Beverly Hills

Miller, D. (1988), 'Appropriating the state on the council estate', *Man*, vol. 23, pp. 353–72

Mills, C.W. (1951), *White Collar*, Oxford University Press, New York

Mills, C.W. (1959), *The Sociological Imagination*, Oxford University Press, New York

Miner, H. (1973), 'Body Ritual among the Nacirema', in D.R. MacQueen, *Understanding Sociology through Research*, Addison-Wesley, Reading, Massachusetts

Minichiello, V., Aroni, R., Timewell, E. and Alexander, L. (1990), *In-depth Interviewing: Researching People*, Longman Cheshire, Melbourne

Mounin, G. (1985), *Semiotic Praxis*, Plenum, New York

Murray, K.D.S. (1992), *The Judgment of Paris: Recent French Theory in a Local Context*, Allen & Unwin, Sydney

Myers, D. (1991), 'On, onward to battle' (Letters), *The Australian* (Higher Education Supplement), 18 December, p. 18

Najman, J., Morrison, J., Williams, G.M. and Anderson, M.J. (1992), 'Comparing alternative methodologies in social research: an overview', in J. Daly, I. McDonald and E. Willis (eds), *Researching Health Care: Designs, Dilemmas and Disciplines*, Routledge, London

Nash, J.E. (1989), 'What's in a face? The social character of the English bulldog', *Qualitative Sociology*, vol. 12, no. 4, pp. 357–67

Neelamkavil, F. (1987), *Computer Simulation and Modelling*, John Wiley, Chichester

Nettl, B. (1956), *Music in Primitive Culture*, Harvard University Press, Cambridge

Nettl, B. (1983), *The Study of Ethnomusicology: Twenty-nine Issues and Concepts*, University of Illinois Press, Urbana

Norris, C. (1990), *What's Wrong with Postmodernism*, Harvester Wheatsheaf, New York

Palca, D. (1981), *The Language of Clothes*, Random House, New York

Pattison, R. (1987), *The Triumph of Vulgarity*, Oxford University Press, New York

Patton, M.Q. (1980), *Qualitative Evaluation Methods*, Sage, Beverly Hills

Pederson, A. (1987), *Keeping Archives*, Australian Society of Archivists Inc., Sydney

Perkins, J. (1992), *The Treasurer Game*, Economic Communications, Melbourne

Perry, J. (ed.) (1989), *Doing Fieldwork: Eight Personal Accounts of Social Research*, Deakin University Press, Geelong

Petersen, R.A. and Berger, D.G. (1975), 'Cycles in symbol production: the case of popular music', *American Sociological Review*, vol. 40 (April) pp. 158–73

Plummer, K. (1983), *Documents of Life*, George Allen & Unwin, London

Rathje, W.L. (1979), 'Trace measures', in L. Sechrest (ed.), *Unobtrusive Measures Today*, Jossey-Bass, San Francisco

Richards, L. and Richards, T. (1991), 'Computing in qualitative analysis: a healthy development?', *Qualitative Health Research*, vol. 1, no. 2, pp. 234–62

Ricketts, L.M. (1978), 'Caesar's army in Gaul: a computer-aided prosopography', in P. C. Patton and R.A. Holoien (eds), *Computing in the Humanities*, Gower, pp. 197–204

Ricoeur, P. (1971), 'The model of the text: meaningful action considered as a text', *Social Research*, vol. 38, pp. 529–62

Robinson, J. and Hirsch, P. (1972), 'Teenage response to rock and roll protest songs', in R.S. Denisoff and R.A. Peterson (eds), *The Sounds of Social Change*, Rand McNally, Chicago

Robinson, P. (1991), 'Big bins more efficient: what a load of rubbish', *The Sunday Age*, 24 November, News p. 5

Rollwagen, J.R. (1988), *Anthropological Filmmaking*, Harwood Academic Publishers, London

Rotheschild, N.A. (1981), 'Pennies from Denver', in R.A. Gould and M.B. Schiffer (eds), *Modern Material Culture: The Archaeology of Us*, Academic Press, New York, pp. 161–81

Roughley, A. (1991), 'Deconstruction revisted: a constructive exploration', *The Australian* (Higher Education Supplement), p. 22

Runcie, J.F. (1980), *Experiencing Social Research*, Dorsey Press, Illinois

Saussure de, F. (1966), *Course in General Linguistics*, McGraw-Hill, New York

Savoury, L. and O'Connor, T. (1973), *The Heart Has Its Seasons: Reflections on the Human Condition*, Regina Press, New York

Schatzman, L. and Strauss, A.L. (1973), *Field Research: Strategies for a Natural Sociology*, Prentice-Hall, Englewood Cliffs, New Jersey

Schwartz, D. (1989), 'Visual ethnography: using photography in qualitative research', *Qualitative Sociology*, vol. 12, no. 2, pp. 119–54

Schwartz, H. and Jacobs, J. (1979), *Qualitative Sociology: A Method to the Madness*, The Free Press, New York

Schwirtlich, A. (1987), 'Introducing archives and the archival profession', in A. Pederson (ed.), *Keeping Archives*, Australian Society of Archivists Inc., Sydney, pp. 1–20

Scott, J. (1990), *A Matter of Record*, Polity Press, Cambridge

Sechrest, L. and Phillips, M. (1979), 'Unobtrusive measures: an overview', in L. Sechrest (ed.), *Unobtrusive Measures Today*, Jossey-Bass, San Francisco

Sechrest, L. (ed.) (1979), *Unobtrusive Measures Today*, Jossey-Bass, San Francisco

Seiter, E. (1987), 'Semiotics and television', in R.C. Allen (ed.), *Channels of Discourse*, Routledge, London

Shepherd, J. (1987), 'Towards a sociology of musical styles', in A.L. White (ed.), *Lost in Music*, Routledge & Kegan Paul, London

Sherman, B.L. and Dominick, J.R. (1986), 'Violence and sex in music videos: T.V. and rock 'n' roll, *Journal of Communication*, vol. 36, no. 1, pp. 79–93

SIM CITY (1989–91), (For MacIntosh Apple Users), Maxis Software, Orinda, California

Singleton, R., Straights, B.C., Straights, M.M. and McAllister, R.J. (1988), *Approaches to Social Research*, Oxford University Press, New York

Smith, K. (1981), *Newstalk: Women's Place in the Language of Adelaide Newspapers*, Occasional Papers in Language, Media and Culture. Department of Education, University of Adelaide

Social Sciences Data Archives (1991), *SSDA Catalogue*, Australian National University, Canberra

Sontag, S. (1979), *On Photography*, Penguin, Harmondsworth

Spooner, B. (1986), 'Weavers and dealers: the authenticity of an oriental carpet', in A. Appadurai (ed.), *The Social Life of Things*, Cambridge University Press, Cambridge

Spy Hole (1992), *The Sunday Age*, 8 March

Stanfield, J. (1987), 'Archival methods in race relations research', *American Behavioural Scientist*, vol. 30, no. 4, pp. 366–80

Stefani, G. (1973), 'Séiotique en musicologie', *Versus*, vol. 5, pp. 20–42

Stewart, D. (1984), *Secondary Research: Information, Sources and Methods*, Sage, Beverly Hills

Strauss, A.L. (1987), *Qualitative Analysis for Social Scientists*, Cambridge University Press, Cambridge

Thomas, G.V. and Silk, A. (1990), *An Introduction to the Psychology of Children's Drawings*, Harvester Wheatsheaf, London

Thomas, W. and Zuaniecki, F. (1958), *The Polish Peasant in Europe and America*, Dover, New York

Thompkins, E. (1991), 'Deconstruction's halo slips' (Letters), *The Australian* (Higher Education Supplement)

Turner, R. (1989), 'Deconstructing the field', in J.F. Gurbrium and D. Silverman (eds), *The Politics of Field Research*, Sage, London, pp. 13–30

Vera, H. (1989), 'On Dutch windows', *Qualitative Sociology*, vol. 12, no. 2, pp. 215–34

Viney, L.L. and Bousefield, L. (1991), 'Narrative analysis: a method of psychosocial research for AIDS affected people', *Social Science and Medicine*, vol. 32, no. 7, pp. 757–65

Wadsworth, Y. (1984), *Do It Yourself Social Research*, Victorian Council of Social Services, Collingwood, Victoria

Walker, S. (ed.) (1983), *Who is She? Images of Women in Australian Fiction*, University of Queensland Press, St Lucia

Walker, R. (1991), 'Finding a silent voice for the researcher: using photographs in evaluation and research', in M. Schratz (ed.), *Qualitative Voices in Educational Research*, The Falmer Press, London

Walker, A.L. and Moulton, R.K. (1989), 'Photo albums: images of time and reflections of self', *Qualitative Sociology*, vol. 12, no. 2, pp. 155–83

Walsh, V.A. (182), 'Computer simulation methodology for archaeology', *Computing in the Humanities*, vol. 5, pp. 163–74

Webb, E., Campbell, D., Schwartz, R. and Sechrest, L. (1966), *Unobtrusive Measures: Non-Reactive Research in the Social Sciences*, Rand McNally & Co., Chicago

Webb, E., Campbell, D., Schwarts, R., Sechrest, L. and Grove, J. (1981), *Non-Reactive Measures in the Social Sciences*, Houghton Mifflin Co., Boston

Weber, R.P. (1990), *Basic Content Analysis*, Sage, Newbury Park

Werner, O. and Schoepfle, G.M. (1987) *Systematic Fieldwork Vols. 1 and 2*, Sage, Newbury Park

Wernick, A. (1991), *Promotional Culture: Advertising, Ideology and Symbolic Expression*, Sage, London

Whicker, M.L. and Sigelman, L. (1991), *Computer Simulation Applications*, Sage, Newbury Park

Williamson, J. (1986), *Consuming Passions*, Marion Boyers, London

Willis, P. E. (1978), *Profane Culture*, Routledge & Kegan Paul, London

Willis, E. (1986), 'Commentary: RSI as a social process', *Community Health Studies*, vol. 10, no. 2, pp. 210–19

Wiseman, J.P. and Aron, M.S. (1970), *Field Projects for Sociology Students*, Schenkman, Cambridge Massachusetts

Wolfenstein, M. and Leites, N. (1950), *Movies: A Psychological Study*, Hafner, New York

Wollen, P. (1986), 'Ways of thinking about music video', *Critical Quarterly*, vol. 28, nos. 1 and 2, pp. 167–170

Woodland, D.J.A. (1979), 'Empirical; empiricism', in G.D. Mitchell, *A New Dictionary of Sociology*, Routledge & Kegan Paul, London, pp. 65–6

Yule, V. (1987), 'Observing adult–child interaction: an example of a piece of research anyone could do', in M. O'Connell (ed.), *New Introductory Reader in Sociology*, Nelson, Edinburgh, pp. 69–75

Zaetta, L. (1980), *For Christ's Sake*, Unicorn, East Melbourne

Ziller, R.C. (1990), *Photographing the Self: Methods for Observing Personal Orientations*, Sage, Newbury Park

Zolberg, V.L. (1983), *Constructing a Sociology of the Arts*, Cambridge University Press, Cambridge

Index

SURVEYS IN SOCIAL RESEARCH
Third Edition

by David de Vaus

This third and substantially revised edition of *Surveys in Social Research* provides clear advice on how to plan, conduct and analyse social surveys. It emphasises the links between theory and research, the logic and interpretation of statistics and the process of social research.

David de Vaus shows readers how to both conduct good surveys and to become critical consumers of research. He stresses that the logic of surveys and statistics is simply an extension of the logic we use in everyday life, but that analysis is an art that requires creativity and imagination rather than the application of mechanical procedures. As a consequence, he highlights the plain logic, selection and interpretation of statistics rather than formulae and computation.

This third edition includes substantial new sections on question design and pilot testing of questionnaires, factor analysis and coding. Useful additions have also been made to further clarify techniques and to warn researchers of problems that may arise. References have been fully updated and some new exercises provided.

Surveys in Social Research is essential reading for all students in the social sciences. It is written in the knowledge that many social science students have little background in statistical analysis, and it aims to alleviate the trepidation with which they enter this field.

THE SOCIOLOGICAL QUEST
An Introduction to the Study of Social Life

by Evan Willis

This user-friendly introduction to sociology takes the reader on a quest, guiding them along the path towards a sociological understanding of the world in which we live.

Using contemporary examples, *The Sociological Quest* asks what is distinctive about the way sociologists view society. Evan Willis argues that they are concerned with the relationship between the individual and society and that a sociological analysis involves an approach that is historical, cultural, structural and critical.

The Sociological Quest is for anyone thinking of undertaking a study of sociology, as well as for sociology students looking for a concise and useful introduction to the subject.